ECHO SPOT 2024 USER GUIDE

Master Smart Home Integration, Enhance Daily Routines and Boost Home Security

Michael S. Nilsson

Copyright © 2024 by Michael S. Nilsson.

All rights reserved. No part of this publication may be reproduced, distributed, or transmitted in any form or by any means, including photocopying, recording, or other electronic or mechanical methods, without the prior written permission of the author, except in the case of brief quotations embodied in critical reviews and certain other non-commercial uses permitted by copyright law.

TABLE OF CONTENTS

Introduction 6
Chapter 1 12
The Evolution of Smart Speakers 12
    History of Amazon Echo Devices 12
    Key Milestones and Innovations 13
    Transition from Simple Speakers to Multifunctional Smart Devices 14
    Overview of the Echo Spot's Predecessors and Their Impact 15
Chapter 2 18
Unboxing and First Impressions 18
    Detailed Description of the Echo Spot 2024 Packaging 18
    Step-by-Step Unboxing Guide with Images 19
    Initial Setup Process and First Impressions 21
    Comparison with Previous Echo Spot Models and Other Amazon Devices 22
Chapter 3 28
Detailed Features and Specifications 28
    Comprehensive Breakdown of the Echo Spot's Hardware 28
    Touchscreen Functionality and Display Specifications.. 30
    Speaker Quality and Audio Features 31
    Connectivity Options: Wi-Fi, Bluetooth, and More 33
    Matter Controller Capabilities and Smart Home Integration 34

New and Unique Features Introduced in the 2024 Model..35

Chapter 4..**40**
Setting Up Your Echo Spot...................................**40**
Step-by-Step Guide to Setting Up the Echo Spot......40
Connecting to Wi-Fi and Bluetooth........................42
Integrating with the Alexa App..............................43
Customizing Settings for Optimal Use...................44
Tips for a Seamless Setup Experience..................46

Chapter 5..**48**
Smart Home Integration.......................................**48**
Overview of Smart Home Ecosystems...................48
How to Connect the Echo Spot with Various Smart Home Devices..49
Using Alexa for Home Automation.........................50
Creating and Managing Routines for Different Scenarios..51
Real-World Examples of Smart Home Integration.....53
Advanced Smart Home Integration........................55
Optimizing Your Smart Home for Efficiency and Convenience...57

Chapter 6..**60**
Daily Routines and Productivity..........................**60**
Using the Echo Spot to Enhance Daily Routines......60
Setting Alarms, Reminders, and Calendar Events....62
Utilizing Alexa for Productivity Tasks.....................63
Integrating with Third-Party Productivity Apps...........64
Case Studies of Improved Productivity Using Echo Spot...66
Tips for Maximizing Productivity with the Echo Spot. 68

3

Chapter 7 .... **72**
Entertainment and Media **72**
    Using Echo Spot for Music and Audiobooks 72
    Streaming Services Compatibility and Setup 74
    Using the Touchscreen for Media Control 75
    Overview of Kindle Read-Aloud and Other Unique Media Features 76
    Personalizing Entertainment Settings for an Optimal Experience 78

Chapter 8 .... **82**
Privacy and Security **82**
    Understanding Privacy Concerns with Smart Devices .. 82
    Built-in Privacy Features of the Echo Spot 83
    Managing Microphone and Camera Settings 85
    Ensuring Secure Connections and Data Protection .. 86
    Best Practices for Maintaining Privacy and Security. 88

Chapter 9 .... **92**
Advanced Tips and Customizations **92**
    Advanced Settings and Customizations for Power Users 92
    Personalizing the Display and Touch Interface 93
    Customizing Alexa Responses and Interactions 95
    Using the Echo Spot in Unique and Creative Ways .. 96
    Integration with Other Amazon Services and Devices ... 98

Chapter 10 .... **102**
Troubleshooting and Support **102**
    Common Issues and How to Resolve Them 102
    Step-by-Step Troubleshooting Guides 104

When and How to Contact Amazon Support........... 105
Accessing Community Resources and Forums for Additional Help...106
Maintaining and Updating Your Echo Spot............. 107

Chapter 11.. 110
Future of Smart Home Technology..................................110
Emerging Trends in Smart Home Technology......... 110
Future Developments in Amazon Echo Devices......112
How the Echo Spot Fits into the Evolving Smart Home Landscape... 114
Predictions and Expectations for the Next Generation of Smart Speakers.. 115

Conclusion..120
Recap of Key Points Covered in the Book...............120
Final Thoughts on Maximizing the Benefits of the Echo Spot.. 123
Encouragement to Continue Exploring Smart Home Technology..125

Appendices... 130
Glossary of Terms..130
Frequently Asked Questions (FAQ)......................... 131
Additional Resources for Further Learning.............. 133
List of Compatible Devices and Services................ 134
Detailed Specifications and Technical Information.. 135

About the Author... 140

5

INTRODUCTION

Welcome to a world where technology seamlessly integrates into our daily lives, making our homes smarter, our routines more efficient, and our interactions with devices almost magical. Imagine waking up to a soothing voice that not only tells you the weather but also plays your favorite morning playlist, reads your Kindle books aloud while you sip your coffee, and even prepares your home for the day ahead. Welcome to the future of smart home living, embodied in the Amazon Echo Spot 2024.

The Evolution of Smart Home Technology

The concept of a smart home has evolved dramatically over the past decade. What once seemed like science fiction is now a reality, thanks to rapid advancements in technology. From the first clunky home automation systems to today's sleek and intuitive smart devices, the journey has been nothing short of extraordinary. Smart home technology began with simple, standalone devices like programmable thermostats and remote-controlled lights. These early innovations paved the way for a more connected and integrated approach to home automation. Today, we live in an era where almost every aspect of our homes can be controlled with a simple voice command or a tap on our smartphones. This transformation has been driven by a relentless pursuit of convenience, efficiency, and enhanced quality of life. As the technology matured, so did our expectations. We now demand more than just basic functionality; we seek devices that are intelligent, adaptive, and capable of learning our preferences.

The Importance and Benefits of Smart Speakers

At the heart of this smart home revolution lies the smart speaker. These unassuming devices have redefined the way we interact with technology. Smart speakers like the Amazon Echo series have become indispensable companions, offering a myriad of benefits that go beyond just playing music or setting timers. They have become our personal assistants, home automation hubs, and even sources of entertainment and education.

The benefits of smart speakers are manifold:

- Convenience: With voice commands, you can control various aspects of your home without lifting a finger. Whether it's adjusting the thermostat, turning off the lights, or playing your favorite song, smart speakers make it effortless.
- Efficiency: Smart speakers streamline your daily routines. They can remind you of appointments,

help you manage your to-do lists, and even order groceries for you.
- Entertainment: From streaming music and audiobooks to providing real-time news updates and trivia games, smart speakers keep you entertained and informed.
- Integration: As the central hub of your smart home, these devices connect and control a wide range of compatible smart home products, creating a cohesive and synchronized environment.

Introduction to the Echo Spot 2024 and Its Relevance

Enter the Amazon Echo Spot 2024, the latest innovation from Amazon that promises to elevate your smart home experience to new heights. Building on the success of its predecessors, the Echo Spot 2024 is not just an incremental upgrade; it's a reimagining of what a smart speaker can be. Designed with a sleek, compact form factor, it seamlessly fits into any room, especially your bedside table.

The Echo Spot 2024 is a game-changer in many ways. It boasts a vibrant 2.83-inch touchscreen, enhanced audio capabilities, and an array of new features tailored for a more intuitive and interactive user experience. Whether you're a tech enthusiast looking to stay ahead of the curve or someone seeking to simplify their daily life, the Echo Spot 2024 offers something for everyone.

One of the standout features of the Echo Spot 2024 is its versatility. It serves as an alarm clock, media player, and smart home controller, all rolled into one compact device. Imagine waking up to a gentle alarm, followed by a personalized briefing of your day, complete with weather updates, calendar events, and news highlights. As you go about your morning routine, the Echo Spot 2024 keeps you connected and informed, all with just a few voice commands or taps on its touchscreen.

Objectives of the Book and What Readers Can Expect to Learn

This book is your ultimate guide to mastering the Amazon Echo Spot 2024. Whether you're a seasoned smart home user or a newcomer to the world of smart speakers, you'll find valuable insights and practical advice to make the most of your device. Here's what you can expect to learn:

- Comprehensive Setup Guide: We'll walk you through the initial setup process, ensuring that your Echo Spot 2024 is configured perfectly to suit your needs.
- In-Depth Feature Exploration: Discover the full range of features that the Echo Spot 2024 offers,

from basic functionalities to advanced settings and customizations.
- Smart Home Integration: Learn how to seamlessly connect and control your other smart home devices, creating a harmonious and efficient living environment.
- Daily Routines and Productivity: Optimize your daily routines with tips and tricks for using the Echo Spot 2024 as your personal assistant.
- Entertainment and Media: Unlock the full potential of your Echo Spot 2024 as a media hub, enhancing your entertainment experience.
- Privacy and Security: Understand the privacy features and settings to ensure your personal information is secure.
- Troubleshooting and Support: Gain insights into common issues and how to resolve them quickly and efficiently.
- Future Trends: Get a glimpse of what's next in the world of smart home technology and how the Echo Spot 2024 fits into the broader picture.

By the end of this book, you'll not only be an Echo Spot 2024 expert but also a savvy smart home enthusiast, ready to embrace the future of connected living. So, let's embark on this journey together and unlock the full potential of your smart home with the Amazon Echo Spot 2024.

CHAPTER I

THE EVOLUTION OF SMART SPEAKERS

The journey of smart speakers has been nothing short of revolutionary, transforming the way we interact with our homes and the technology around us. At the forefront of this revolution is the Amazon Echo series, a lineup of devices that have redefined convenience and automation in our daily lives. This chapter delves into the history of Amazon Echo devices, key milestones in their development, and the transition from simple speakers to multifunctional smart devices. We'll also explore the predecessors of the Echo Spot and their impact on the smart home ecosystem.

History of Amazon Echo Devices

It all began in November 2014 when Amazon introduced the first-generation Echo. This cylindrical device, equipped with seven microphones and a powerful speaker, was more than just a novelty. It brought Alexa, Amazon's voice-activated assistant, into homes, allowing users to control music playback, ask questions, and manage their schedules with simple voice commands.

The initial skepticism soon gave way to widespread adoption as people discovered the convenience and capabilities of this new technology.

The first Echo set the stage for a series of innovations. In 2016, Amazon released the Echo Dot, a smaller and more affordable version that could be connected to external speakers. This move made Alexa more accessible and integrated into more aspects of users' lives. The Echo Dot quickly became a best-seller, solidifying Amazon's position in the smart speaker market.

Key Milestones and Innovations

As the Echo lineup expanded, so did its capabilities. The introduction of the Echo Show in 2017 marked a significant milestone. With its built-in screen, the Echo Show combined voice control with visual information, enabling video calls, streaming, and displaying recipes, weather updates, and more. This was followed by the Echo Spot in 2018, which offered a compact design with a smaller screen, ideal for bedside tables and kitchens.

One of the most significant innovations came with the Echo Plus, which featured a built-in smart home hub. This allowed the device to directly connect and control a wide range of smart home devices, simplifying the setup and integration process for users. Amazon continued to enhance Alexa's capabilities, introducing features like multi-room audio, voice recognition, and routines, which allowed users to automate a series of actions with a single command.

Transition from Simple Speakers to Multifunctional Smart Devices

The evolution of the Echo devices is a reflection of the broader transition from simple speakers to multifunctional smart devices. Initially, the primary function of the Echo was to play music and provide information upon request. However, as the ecosystem grew, these devices became central hubs for smart home automation.

This transition was driven by continuous improvements in artificial intelligence and natural language processing, which made Alexa more intuitive and responsive. The integration of third-party skills and services further extended the functionality of Echo devices, allowing users to order groceries, book rides, control home security systems, and much more, all through voice commands.

The expansion of the Echo lineup to include devices like the Echo Auto for cars and the Echo Frames, which integrate Alexa into eyeglasses, demonstrates Amazon's vision of making Alexa a ubiquitous presence in users' lives. These innovations have transformed the way we interact with technology, making voice control a natural and essential part of our daily routines.

Overview of the Echo Spot's Predecessors and Their Impact

Before the Echo Spot, several key devices laid the groundwork for its development. The Echo Show's introduction of a screen was a game-changer, adding a visual dimension to Alexa's capabilities. The success of the Echo Show proved that there was a demand for smart speakers with displays, leading to the creation of more compact and versatile devices like the Echo Spot.

The Echo Dot's affordability and versatility played a crucial role in popularizing Alexa. By making it easy to place Alexa in multiple rooms, the Echo Dot helped users experience the benefits of a connected home. This widespread adoption created a market that was ready for more specialized devices like the Echo Spot.

The Echo Plus's integration of a smart home hub simplified the user experience, making it easier for people to adopt and manage smart home technologies. This innovation laid the foundation for the Echo Spot to serve not only as a smart speaker but also as a control center for other smart devices in the home.

The impact of these predecessors on the Echo Spot is evident in its design and functionality. The Echo Spot combines the visual capabilities of the Echo Show with the compact form factor of the Echo Dot, while also incorporating the smart home integration features of the Echo Plus. This combination makes the Echo Spot a versatile and powerful device that enhances the smart home experience in a unique way.

The evolution of the Amazon Echo series has been marked by continuous innovation and a deep understanding of user needs. From the first-generation Echo to the latest Echo Spot, each device has contributed to the development of a more connected and intuitive smart home ecosystem. As we continue to embrace these technologies, the possibilities for enhancing our daily lives with smart speakers are boundless.

CHAPTER 2

UNBOXING AND FIRST IMPRESSIONS

The excitement of unboxing a new gadget is always a unique experience, and the Echo Spot 2024 is no exception. From the moment you lay your hands on the sleek, well-designed packaging, you know you're in for something special. This chapter takes you through the unboxing process, the initial setup, and provides a comparison with its predecessors and other Amazon devices.

Detailed Description of the Echo Spot 2024 Packaging

The Echo Spot 2024 arrives in a compact, minimalist box that immediately signals modernity and sophistication. The packaging is adorned with high-resolution images of the device, showcasing its sleek design and highlighting key features. The front of the box prominently displays the Echo Spot itself, while the sides offer a brief overview of its capabilities, including voice control, touchscreen interface, and smart home integration.

As you turn the box around, you'll find detailed specifications and a list of what's included in the package. The materials used for the packaging are eco-friendly, aligning with Amazon's commitment to sustainability. The box feels sturdy and protective, ensuring that the Echo Spot reaches you in pristine condition.

Step-by-Step Unboxing Guide with Images

Upon opening the box, you're greeted with a quick start guide that sits atop a protective foam layer. Removing the foam reveals the Echo Spot 2024, nestled securely in a custom-fit compartment. The first impression of the device is its compactness and elegance, with a smooth, rounded form factor that is both modern and practical.

To the side of the Echo Spot, you'll find the power adapter neatly tucked into its own compartment, along with a small packet containing warranty information and safety instructions. The power adapter is designed to be slim and unobtrusive, making it easy to fit into any power outlet without blocking adjacent plugs.

Here's a step-by-step guide to unboxing your Echo Spot 2024:

1. *Open the Box:* Carefully cut the seal and open the box to reveal the quick start guide.
2. *Remove the Foam Layer:* Lift out the protective foam layer to access the Echo Spot.

3. Take Out the Echo Spot: Gently lift the Echo Spot from its compartment.

4. Unpack the Power Adapter: Remove the power adapter from its compartment.

5. Check the Documentation: Take out the warranty and safety information packet.

The unboxing process is straightforward and user-friendly, ensuring that you can get started with your Echo Spot quickly and without hassle.

Initial Setup Process and First Impressions

Setting up the Echo Spot 2024 is designed to be a seamless experience, even for those who are not tech-savvy. Here's how you can get your device up and running in no time:

1. Plug In the Device: Connect the power adapter to the Echo Spot and plug it into an outlet. The device will power on automatically.

2. Follow On-Screen Instructions: The Echo Spot's touchscreen will display a series of prompts to guide you through the setup process. This includes selecting your preferred language, connecting to Wi-Fi, and logging into your Amazon account.

3. Configure Alexa: Once connected, you'll be prompted to set up Alexa. This involves configuring voice recognition and linking any existing smart home devices.
4. Customize Settings: Adjust settings such as display brightness, sound preferences, and privacy options according to your preferences.

The initial setup is intuitive, with clear on-screen instructions that make the process effortless. Within minutes, your Echo Spot will be ready to use, and you'll be greeted by Alexa, ready to assist you with a myriad of tasks.

First impressions of the Echo Spot 2024 are overwhelmingly positive. The device's design is both aesthetically pleasing and functional, fitting seamlessly into any room decor. The touchscreen is responsive and clear, providing a user-friendly interface for interacting with Alexa and accessing various features. The sound quality is impressive for such a compact device, delivering clear and balanced audio that is perfect for both music and voice responses.

Comparison with Previous Echo Spot Models and Other Amazon Devices

The Echo Spot 2024 stands out not only for its new features but also for the enhancements it brings compared to previous models and other Amazon devices. Here's a detailed comparison:

Echo Spot (2018) vs. Echo Spot 2024:

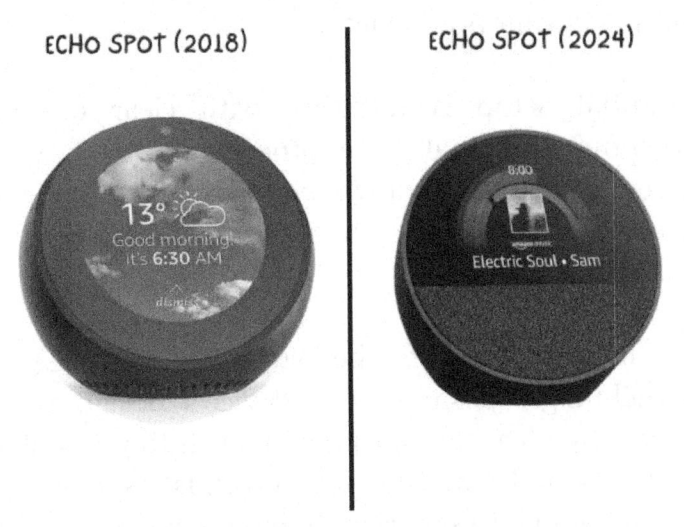

ECHO SPOT (2018) | ECHO SPOT (2024)

- Design: The 2024 model retains the compact, spherical design but features a slightly larger and higher resolution touchscreen, enhancing visual clarity and touch responsiveness.
- Audio Quality: Improvements in speaker technology provide richer and clearer audio, making it a better choice for both music and voice commands.
- Smart Home Integration: While the 2018 model offered basic smart home integration, the 2024 version comes with enhanced compatibility and a

more intuitive setup process, making it easier to connect and control various smart home devices.
- Privacy Features: The Echo Spot 2024 includes updated privacy controls, such as a physical microphone off button and more robust data protection measures.

Echo Show Series vs. Echo Spot 2024:

ECHO SHOW SERIES	ECHO SPOT (2024)

- Size and Form Factor: The Echo Show devices, particularly the Echo Show 5 and Echo Show 8, offer larger screens but are also bulkier. The Echo Spot 2024's compact design makes it ideal for bedside tables and small spaces where a larger device might be cumbersome.
- Display Features: While the Echo Show series boasts larger displays suitable for video calls and

streaming, the Echo Spot 2024's screen is optimized for quick, glanceable information, such as weather updates, calendar events, and notifications.
- Functionality: Both the Echo Show and Echo Spot 2024 offer similar smart home control capabilities and access to Alexa's extensive skills library. However, the Echo Spot's design is tailored for users who prioritize space efficiency and aesthetic appeal.

Echo Dot vs. Echo Spot 2024:

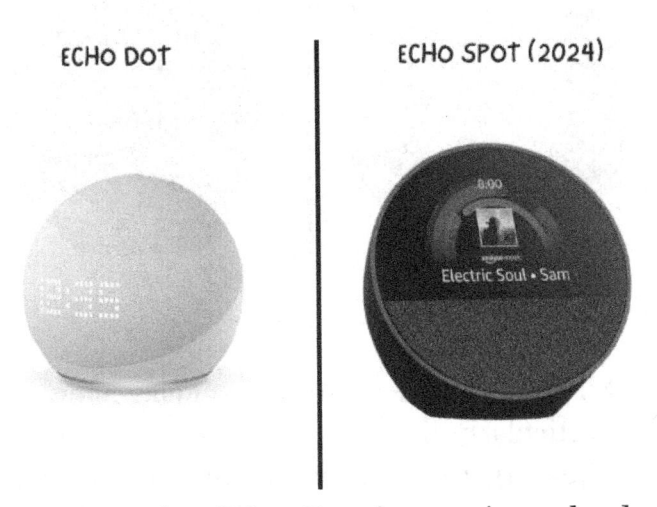

ECHO DOT	ECHO SPOT (2024)

- Screen: The Echo Dot is a voice-only device, whereas the Echo Spot 2024 includes a touchscreen, adding a visual dimension to interactions with Alexa.

- Use Cases: The Echo Dot is perfect for those who need a basic, affordable smart speaker for voice commands and audio playback. In contrast, the Echo Spot 2024 is designed for users who want a multifunctional device that can also display information and serve as a smart alarm clock.

The Echo Spot 2024 emerges as a versatile and powerful addition to the Echo family, combining the best features of its predecessors with new enhancements that cater to modern smart home needs. Its compact design, improved audio and display capabilities, and seamless smart home integration make it an ideal choice for users looking to elevate their smart home experience.

As we move forward, the Echo Spot 2024 promises to be a central hub for convenience, productivity, and entertainment, all while fitting elegantly into the aesthetic of any room. Whether you're new to smart speakers or an experienced user, the Echo Spot 2024 offers a blend of simplicity and sophistication that is sure to impress.

CHAPTER 3

DETAILED FEATURES AND SPECIFICATIONS

As we delve deeper into the Echo Spot 2024, it's time to explore the intricate details that make this device a standout in the world of smart home technology. From its robust hardware to the advanced capabilities that set it apart, this chapter provides a comprehensive breakdown of everything you need to know about the Echo Spot 2024.

Comprehensive Breakdown of the Echo Spot's Hardware

The Echo Spot 2024 is a marvel of modern engineering, blending sleek design with powerful functionality. Its compact, spherical form factor is both aesthetically pleasing and highly practical, making it a versatile addition to any room in your home.

Design and Build:
The Echo Spot 2024 features a rounded, minimalist design that is both modern and timeless. The device

measures just under five inches in diameter, making it compact enough to fit on a bedside table or a kitchen counter without taking up too much space. The exterior is made of high-quality plastic with a matte finish, available in three elegant colors: charcoal, glacier white, and twilight blue.

Touchscreen Display:
The upper half of the Echo Spot 2024 is dominated by a vibrant 2.83-inch touchscreen. This display is not only functional but also a key aesthetic feature, blending seamlessly with the device's spherical design. The screen has a resolution of 240x320 pixels, ensuring that text and images are crisp and clear. The touchscreen is highly responsive, making it easy to navigate through menus and interact with various features.

Speaker and Microphone Array:
The Echo Spot 2024 is equipped with a front-firing 1.73-inch speaker, delivering clear and balanced audio. This speaker is designed to handle both music playback and voice interactions with Alexa. The device also features a far-field microphone array, allowing it to pick up voice commands from across the room, even in noisy environments.

Buttons and Ports:
On the back of the Echo Spot 2024, you'll find three physical buttons: volume up, volume down, and a

microphone mute button. These buttons provide easy access to essential controls without needing to navigate through menus. The device also includes a power port and a 3.5mm audio output for connecting to external speakers.

Touchscreen Functionality and Display Specifications

The touchscreen is one of the standout features of the Echo Spot 2024, adding a visual dimension to the device's functionality. Here's a closer look at what the screen offers:

Interactive Display:
The 2.83-inch touchscreen is designed for intuitive interaction. Users can swipe through different screens, tap to select options, and even pinch to zoom on certain content. This makes it easy to navigate through the device's features and customize settings on the fly.

Visual Information:
The screen displays a variety of useful information, including the time, weather forecasts, calendar events, and notifications. This makes the Echo Spot 2024 an ideal bedside companion, providing all the essential information you need at a glance.

Customizable Faces:
One of the fun aspects of the touchscreen is the ability to customize the clock face. Users can choose from a range of designs, including analog and digital options, as well as themed faces that add a personal touch to the device.

Media Playback:
The touchscreen also enhances media playback. Users can see album art while listening to music, view video thumbnails, and control playback with on-screen buttons. This visual feedback adds an extra layer of engagement to the listening experience.

Smart Home Control:
Through the touchscreen, users can easily control smart home devices connected to Alexa. The screen provides a visual interface for adjusting settings, such as dimming lights or changing thermostat settings, making home automation more intuitive.

Speaker Quality and Audio Features

Despite its compact size, the Echo Spot 2024 delivers impressive audio performance, thanks to its well-engineered speaker system.

Front-Firing Speaker:

The device's front-firing speaker is designed to provide clear and balanced sound, whether you're listening to music, engaging in a conversation with Alexa, or following along with an audiobook. The speaker is optimized to handle a wide range of frequencies, ensuring that both high and low tones are reproduced accurately.

Enhanced Audio Processing:
The Echo Spot 2024 features advanced audio processing algorithms that enhance sound quality. This includes noise reduction and echo cancellation, which ensure that Alexa can hear your commands clearly, even when music is playing.

Bluetooth and 3.5mm Audio Output:
For those who prefer a more powerful audio setup, the Echo Spot 2024 includes Bluetooth connectivity, allowing you to pair it with external speakers. Additionally, the 3.5mm audio output provides a wired option for connecting to home stereo systems.

Voice Recognition:
The far-field microphone array ensures that the Echo Spot 2024 can pick up voice commands from across the room, even in noisy environments. The device is designed to recognize and respond to voices accurately, making it a reliable assistant for any household.

Connectivity Options: Wi-Fi, Bluetooth, and More

Connectivity is a critical aspect of any smart device, and the Echo Spot 2024 excels in this area with robust options that ensure seamless integration into your home network.

Wi-Fi:
The Echo Spot 2024 supports dual-band Wi-Fi (2.4GHz and 5GHz), providing a stable and fast connection for all its online functions. This ensures smooth streaming of music and videos, as well as quick responses to voice commands.

Bluetooth:
In addition to Wi-Fi, the Echo Spot 2024 includes Bluetooth connectivity. This allows you to pair the device with other Bluetooth-enabled devices, such as external speakers, headphones, or even your smartphone. This flexibility enhances the versatility of the Echo Spot, making it easy to integrate into your existing audio setup.

Matter Controller:
The Echo Spot 2024 includes support for Matter, the new standard for smart home devices. Matter aims to unify various smart home ecosystems, ensuring that devices from different manufacturers can work together

seamlessly. As a Matter controller, the Echo Spot 2024 can manage and interact with a wide range of Matter-compatible devices, making it a central hub for your smart home.

Matter Controller Capabilities and Smart Home Integration

The integration of Matter support is a significant step forward for the Echo Spot 2024, enhancing its role as a smart home hub.

Unified Smart Home Ecosystem:
Matter is designed to simplify the smart home experience by providing a universal standard for device compatibility. With the Echo Spot 2024 serving as a Matter controller, you can connect and control a diverse range of smart home devices, regardless of their brand. This creates a more cohesive and user-friendly smart home environment.

Seamless Device Setup:
The Echo Spot 2024 makes it easy to add new Matter-compatible devices to your smart home setup. The setup process is straightforward and streamlined, reducing the complexity often associated with integrating new gadgets. This ease of use encourages the

adoption of smart home technology, making it accessible to a broader audience.

Enhanced Automation:
Matter support enhances the automation capabilities of the Echo Spot 2024. You can create complex routines and automations that involve multiple devices, all managed through the Echo Spot. For example, you can set up a morning routine that turns on the lights, starts the coffee maker, and plays your favorite news podcast—all with a single command.

Improved Interoperability:
With Matter, interoperability between devices is significantly improved. This means that your smart home devices can work together more effectively, providing a more integrated and responsive experience. The Echo Spot 2024, as a Matter controller, ensures that all your devices communicate seamlessly, enhancing the overall functionality of your smart home.

New and Unique Features Introduced in the 2024 Model

The Echo Spot 2024 is not just an incremental upgrade; it introduces several new features that elevate its functionality and user experience.

Adaptive Brightness:
One of the standout features of the Echo Spot 2024 is its adaptive brightness technology. The device automatically adjusts the screen brightness based on the ambient light in the room. This ensures that the display is always easy to read without being too bright or too dim, enhancing usability in different lighting conditions.

Improved Voice Recognition:
The Echo Spot 2024 includes improved voice recognition capabilities, thanks to enhanced natural language processing algorithms. This allows Alexa to understand and respond to commands more accurately, even in noisy environments or when multiple people are speaking.

Whisper Mode:
Whisper Mode is a new feature that allows you to interact with Alexa in a quieter, more discreet manner. This is particularly useful for nighttime use when you don't want to disturb others. When you whisper to Alexa, she responds in a soft, whispering voice, maintaining the device's functionality without compromising on noise levels.

Kindle Read-Aloud:
For avid readers, the Kindle Read-Aloud feature is a game-changer. You can now listen to your Kindle books directly through the Echo Spot 2024. This feature is

especially useful for those who enjoy audiobooks or want to continue their reading experience while doing other tasks.

Customizable Alarm Tones:
The Echo Spot 2024 offers a range of customizable alarm tones, allowing you to choose the perfect sound to wake up to. From gentle melodies to more energetic tunes, you can set alarms that match your personal preference and help you start your day on the right note.

Enhanced Privacy Features:
In response to growing concerns about privacy, the Echo Spot 2024 includes enhanced privacy controls. This includes a physical microphone mute button, as well as more comprehensive data management options within the Alexa app. Users can review and delete voice recordings, ensuring greater control over their personal information.

Expanded Smart Home Integration:
Building on its predecessors, the Echo Spot 2024 offers expanded integration with a broader range of smart home devices. Whether you're using smart lights, thermostats, security cameras, or other connected gadgets, the Echo Spot 2024 provides a unified platform for controlling them all.

Improved Energy Efficiency:

The Echo Spot 2024 is designed with energy efficiency in mind. The device uses less power during standby mode, helping to reduce overall energy consumption. This makes it not only a powerful addition to your smart home but also an environmentally friendly one.

The Echo Spot 2024 is a testament to Amazon's commitment to innovation and user experience. Its combination of advanced features, seamless smart home integration, and thoughtful design enhancements make it a standout device in the smart speaker market. Whether you're new to smart home technology or a seasoned enthusiast, the Echo Spot 2024 offers a comprehensive suite of features that cater to a wide range of needs and preferences. As you continue to explore and integrate this device into your daily life, you'll discover new ways to enhance convenience, productivity, and entertainment in your smart home.

CHAPTER 4

SETTING UP YOUR ECHO SPOT

Setting up your Echo Spot 2024 is an exciting process that sets the stage for a seamless and enhanced smart home experience. This chapter provides a detailed, step-by-step guide to ensure you get your device up and running smoothly. From connecting to Wi-Fi and Bluetooth to integrating with the Alexa app and customizing settings, this guide covers everything you need to know to make the most of your new smart speaker.

Step-by-Step Guide to Setting Up the Echo Spot

Getting started with the Echo Spot 2024 is straightforward, thanks to Amazon's user-friendly setup process. Follow these steps to ensure a smooth setup experience:

1. Unbox Your Echo Spot: Carefully remove the Echo Spot from its packaging. Make sure you have the device, power adapter, and any included documentation.

2. Plug In the Device: Connect the power adapter to the Echo Spot and plug it into a power outlet. The device will power on automatically, and the Amazon logo will appear on the screen.

3. Choose Your Language: Once the device boots up, you'll be prompted to select your preferred language. Use the touchscreen to choose the appropriate option.

4. Connect to Wi-Fi: The Echo Spot will search for available Wi-Fi networks. Select your network from the list and enter the password using the on-screen keyboard. The device will connect to the network, and you'll be ready to proceed.

5. Sign In to Your Amazon Account: You'll be asked to log in to your Amazon account. Enter your email and password to sign in. If you don't have an account, you can create one directly from the device.

6. Set Up Alexa Voice Recognition: The Echo Spot will guide you through the process of setting up Alexa's voice recognition. Follow the prompts to ensure Alexa can understand your voice commands accurately.

7. Customize Initial Settings: You'll have the option to customize initial settings such as time zone, location, and units of measurement (e.g., Celsius vs. Fahrenheit). Adjust these settings as needed to tailor the device to your preferences.

8. *Complete the Setup:* Once you've configured the initial settings, the Echo Spot will finalize the setup process. Alexa will greet you and provide a brief overview of the device's capabilities.

Connecting to Wi-Fi and Bluetooth

A stable internet connection and versatile connectivity options are essential for maximizing the functionality of your Echo Spot. Here's how to connect your device to Wi-Fi and Bluetooth:

Connecting to Wi-Fi:

1. Open Settings: Swipe down from the top of the screen to access the quick settings menu. Tap the gear icon to open the full settings menu.
2. Select Network: Tap on "Network" to view available Wi-Fi networks.
3. Choose Your Network: Select your Wi-Fi network from the list and enter the password. The Echo Spot will connect to the network, and a confirmation message will appear once the connection is successful.

Connecting to Bluetooth:

1. Open Settings: Access the settings menu by swiping down from the top of the screen and tapping the gear icon.

2. Select Bluetooth: Tap on "Bluetooth" to open the Bluetooth settings.

3. Enable Bluetooth: Ensure Bluetooth is turned on. The Echo Spot will begin searching for nearby Bluetooth devices.

4. Pair with a Device: Select the device you want to pair with from the list of available devices. Follow any on-screen prompts to complete the pairing process. Once paired, you can use the Echo Spot to stream audio to the connected device.

Integrating with the Alexa App

The Alexa app is a powerful tool that enhances the functionality of your Echo Spot. It allows you to manage settings, control connected devices, and customize your Alexa experience. Here's how to integrate your Echo Spot with the Alexa app:

1. Download the Alexa App: If you haven't already, download the Alexa app from the App Store (iOS) or Google Play Store (Android).

2. Open the App: Launch the Alexa app on your smartphone or tablet.

3. Sign In: Log in to your Amazon account within the app. Ensure it's the same account you used to set up the Echo Spot.

4. Add Your Echo Spot: Tap the "Devices" icon at the bottom of the screen, then tap the "+" icon in the top right corner to add a new device. Select "Add Device" and choose "Amazon Echo" from the list. Follow the on-screen instructions to complete the setup.

5. Customize Settings: Once your Echo Spot is added, you can customize various settings within the Alexa app. This includes setting up routines, managing connected devices, and adjusting Alexa's preferences.

Customizing Settings for Optimal Use

Personalizing your Echo Spot's settings can significantly enhance your user experience. Here are some key settings to customize for optimal use:

Display Settings:

1. Adjust Brightness: Swipe down from the top of the screen and tap the brightness icon to adjust the display brightness. You can also enable adaptive brightness, which adjusts the screen based on ambient light.

2. Choose Clock Faces: Go to Settings > Home & Clock > Clock & Photo Display. Select from a variety of clock faces and themes to personalize the look of your Echo Spot.

3. Set Photo Slideshows: In the same menu, you can choose to display personal photos as a slideshow. Connect your Amazon Photos account or select from preloaded options.

Sound Settings:

1. Adjust Volume: Use the physical buttons on the device or swipe down from the top of the screen and tap the volume icon to adjust the sound level.

2. Equalizer Settings: Go to Settings > Sounds > Equalizer to customize the bass, midrange, and treble settings for your preferred audio experience.

Privacy Settings:

1. Manage Voice Recordings: Open the Alexa app and go to Settings > Alexa Privacy > Review Voice History. Here, you can review and delete your voice recordings.

2. Enable Do Not Disturb: To avoid interruptions during specific times, go to Settings > Device Options > Do Not Disturb. Set a schedule or enable the feature manually.

Smart Home Settings:

1. Set Up Smart Home Devices: In the Alexa app, tap the "Devices" icon and select "Add Device" to set up new smart home devices. Follow the on-screen instructions to integrate them with your Echo Spot.

2. *Create Routines:* Go to the Routines section in the Alexa app to create custom routines. For example, you can set a routine to turn off the lights, lock the doors, and play soothing music when you say, "Alexa, goodnight."

Tips for a Seamless Setup Experience

Setting up your Echo Spot can be a breeze with a few handy tips to ensure a seamless experience:

1. *Prepare Your Network:* Ensure your Wi-Fi network is stable and has a strong signal where you plan to place your Echo Spot. Avoid areas with potential interference, such as near microwaves or thick walls.

2. *Have Your Credentials Ready:* Before starting the setup, have your Amazon account login details and Wi-Fi password handy. This will save time and prevent interruptions.

3. *Update the Alexa App:* Make sure you have the latest version of the Alexa app installed on your smartphone or tablet. This ensures compatibility and access to the newest features.

4. *Use Voice Commands:* During setup, try using voice commands to interact with Alexa. This helps familiarize

you with the device and ensures Alexa can recognize your voice.

5. Explore the App: Spend some time exploring the Alexa app after setup. Familiarize yourself with the different settings and features to fully utilize your Echo Spot's capabilities.

6. Personalize Your Device: Take advantage of the customization options to make the Echo Spot truly yours. Set up your favorite clock face, choose your preferred news sources, and create routines that fit your lifestyle.

7. Stay Updated: Regularly check for software updates for your Echo Spot. These updates can improve performance, add new features, and enhance security.

Setting up your Echo Spot 2024 is just the beginning of your journey into a smarter, more connected home. By following these steps and tips, you can ensure a smooth setup process and start enjoying all the benefits your new device has to offer. Whether you're integrating it into an existing smart home ecosystem or using it as your first foray into smart home technology, the Echo Spot 2024 is designed to enhance your daily life with its powerful features and intuitive interface.

CHAPTER 5

SMART HOME INTEGRATION

As smart home technology continues to evolve, integrating devices like the Echo Spot 2024 into your ecosystem can significantly enhance your daily life. This chapter provides an in-depth look at smart home ecosystems, how to connect the Echo Spot with various devices, and how to use Alexa for home automation. Additionally, it covers creating and managing routines for different scenarios and provides real-world examples of smart home integration.

Overview of Smart Home Ecosystems

A smart home ecosystem comprises various interconnected devices that work together to automate and streamline tasks within your home. These devices can include smart speakers, lights, thermostats, security cameras, door locks, and more. The goal of a smart home ecosystem is to provide convenience, efficiency, and enhanced security by allowing these devices to communicate and function cohesively.

At the heart of a smart home ecosystem is a central hub that manages and coordinates the interactions between different devices. The Echo Spot 2024, with its advanced capabilities and Matter support, serves as an ideal central hub, enabling seamless integration and control of a wide range of smart home devices.

How to Connect the Echo Spot with Various Smart Home Devices

Connecting your Echo Spot 2024 with various smart home devices is a straightforward process. Here's how you can get started:

1. Ensure Compatibility: Before connecting devices, ensure they are compatible with Alexa. Most modern smart home devices are designed to work with Amazon's ecosystem, but it's always good to check the product specifications.
2. Power On the Device: Make sure the smart home device you want to connect is powered on and in pairing mode. Refer to the device's user manual for specific instructions on how to enable pairing mode.
3. Open the Alexa App: Launch the Alexa app on your smartphone or tablet. Tap the "Devices" icon at the bottom of the screen.
4. Add a New Device: Tap the "+" icon in the top right corner and select "Add Device." Choose the type of

device you want to add from the list (e.g., light, thermostat, camera).

5. Follow On-Screen Instructions: The app will guide you through the setup process, which typically involves connecting the device to your Wi-Fi network and linking it to your Amazon account.

6. Confirm Connection: Once the device is connected, you'll see a confirmation message in the Alexa app. You can now control the device using voice commands through your Echo Spot or via the app.

Using Alexa for Home Automation

Alexa is a powerful tool for home automation, enabling you to control various aspects of your smart home with simple voice commands. Here are some of the ways you can use Alexa for home automation:

Voice Commands:
- Turn on/off lights: "Alexa, turn on the living room lights."
- Adjust thermostat: "Alexa, set the thermostat to 72 degrees."
- Lock doors: "Alexa, lock the front door."
- Check camera feeds: "Alexa, show the front door camera."

Group Control:

You can create groups in the Alexa app to control multiple devices simultaneously. For example, you can group all the lights in your living room and control them with a single command: "Alexa, turn off the living room lights."

Custom Routines:
Routines are a powerful feature that allows you to automate a series of actions based on a single command or trigger. For example, you can create a bedtime routine that turns off the lights, locks the doors, and adjusts the thermostat when you say, "Alexa, goodnight."

Creating and Managing Routines for Different Scenarios

Creating and managing routines can significantly enhance your smart home experience by automating repetitive tasks and creating a more efficient environment. Here's how to create and manage routines in the Alexa app:

1. Open the Alexa App: Launch the Alexa app on your smartphone or tablet.
2. Navigate to Routines: Tap the "More" icon at the bottom right of the screen, then select "Routines."
3. Create a New Routine: Tap the "+" icon in the top right corner to create a new routine.

4. Set a Trigger: Choose what will trigger the routine. This can be a voice command, a schedule, a device action, or a location.

5. Add Actions: Select the actions you want Alexa to perform when the routine is triggered. You can add multiple actions, such as turning on lights, playing music, adjusting the thermostat, or sending notifications.

6. Save the Routine: Once you've added all the desired actions, give your routine a name and tap "Save."

Example Routines:

Morning Routine:
- Trigger: Schedule (e.g., 7:00 AM)
- Actions: Turn on bedroom lights, play news briefing, adjust thermostat to 70 degrees, start the coffee maker.

Leaving Home Routine:
- Trigger: Voice command ("Alexa, I'm leaving")
- Actions: Turn off all lights, lock doors, adjust thermostat to energy-saving mode, arm security system.

Bedtime Routine:
- Trigger: Voice command ("Alexa, goodnight")
- Actions: Turn off lights, lock doors, adjust thermostat, play sleep sounds.

Real-World Examples of Smart Home Integration

To illustrate the practical benefits of smart home integration, let's look at some real-world examples of how the Echo Spot 2024 can enhance your daily life:

Example 1: Smart Lighting Control

John has installed smart lights throughout his home and connected them to his Echo Spot 2024. He uses voice commands to control the lights, creating a more convenient and energy-efficient environment. For instance, he can say, "Alexa, dim the living room lights to 50%," and the Echo Spot adjusts the lights accordingly. John has also set up routines for different times of the day. In the evening, his "Relax" routine dims the lights, plays soft music, and adjusts the thermostat to a comfortable temperature, creating a perfect ambiance for winding down.

Example 2: Enhanced Home Security

Maria uses her Echo Spot 2024 to enhance home security. She has connected her smart door locks, security cameras, and motion sensors to the device. When she's away, she can use the Alexa app to check live camera feeds and receive alerts if any unusual activity is detected. Maria has also set up a "Night Mode" routine that activates at 10:00 PM, locking all doors, turning on outdoor security lights, and arming the security system.

Example 3: Improved Energy Efficiency

David is focused on making his home more energy-efficient. He has connected his smart thermostat, lights, and appliances to his Echo Spot 2024. Using Alexa, he can monitor and control energy usage throughout his home. David has created an "Energy Saver" routine that activates when he leaves for work, adjusting the thermostat to an energy-saving mode, turning off all unnecessary lights, and switching off standby power to appliances.

Example 4: Seamless Entertainment Experience

Sarah loves hosting movie nights at her home. She has integrated her smart TV, sound system, and lighting with her Echo Spot 2024. By creating a "Movie Night" routine, she can set the perfect atmosphere with a single command. When she says, "Alexa, start movie night," the Echo Spot dims the lights, turns on the TV, and adjusts the sound system to her preferred settings. Sarah can also use voice commands to control playback, adjust volume, and switch between different streaming services.

Example 5: Simplified Daily Routines

Mark and Lisa have a busy household with two young children. They use their Echo Spot 2024 to simplify daily routines and keep the household running smoothly. In the morning, their "Wake Up" routine gradually turns on the lights, plays a wake-up playlist, and provides a weather and traffic update. In the evening, their "Homework Time" routine turns off distractions like the TV and gaming consoles, plays soft background music, and sets a timer for study sessions. These routines help the family stay organized and ensure that tasks are completed efficiently.

Advanced Smart Home Integration

For those looking to take their smart home integration to the next level, the Echo Spot 2024 offers advanced features and capabilities:

Multi-Room Audio:
With multi-room audio, you can synchronize music playback across multiple Echo devices in your home. This allows you to enjoy seamless audio experiences, whether you're moving from room to room or hosting a party. To set up multi-room audio, open the Alexa app,

go to "Devices," select "Multi-Room Music," and create a group with the desired Echo devices.

Voice Profiles:
Alexa can recognize different voices and provide personalized responses based on individual preferences. By setting up voice profiles, each member of your household can receive personalized music recommendations, calendar updates, and more. To create a voice profile, open the Alexa app, go to "Settings," select "Alexa Account," and tap "Recognized Voices."

Drop-In and Announcements:
The Drop-In feature allows you to make instant voice or video calls between Echo devices in your home. This is useful for checking in on family members or communicating across different rooms. Announcements enable you to broadcast a message to all Echo devices simultaneously, making it easy to call everyone for dinner or share important information.

Smart Home Skills:
Alexa's capabilities can be extended through smart home skills, which are essentially apps that add new functionalities. Browse the Alexa Skills Store to find and enable skills for controlling specific smart home devices, accessing additional services, or enhancing your smart home experience.

Optimizing Your Smart Home for Efficiency and Convenience

To fully optimize your smart home, consider the following tips:

1. Plan Your Ecosystem: Think about your daily routines and identify areas where smart home devices can add value. Start with a few key devices and gradually expand your ecosystem as needed.

2. Prioritize Compatibility: Choose devices that are compatible with Alexa and the Echo Spot 2024 to ensure seamless integration and control.

3. Group Devices: Create groups in the Alexa app to organize and control multiple devices simultaneously. This simplifies management and allows for more efficient automation.

4. Experiment with Routines: Test different routines to find what works best for your household. Don't be afraid to adjust and refine routines based on your needs and preferences.

5. Leverage Voice Commands: Take advantage of Alexa's voice commands to control devices, manage routines, and access information quickly and conveniently.

6. Stay Informed: Keep up with the latest updates and features for your Echo Spot 2024 and other smart home devices. Regularly check for software updates and explore new skills to enhance your smart home experience.

By integrating the Echo Spot 2024 into your smart home ecosystem, you can create a more connected, efficient, and enjoyable living environment. Whether you're automating daily tasks, enhancing security, or optimizing energy usage, the Echo Spot 2024 offers a versatile and powerful platform for achieving your smart home goals.

CHAPTER 6

DAILY ROUTINES AND PRODUCTIVITY

The Echo Spot 2024 is more than just a smart speaker; it's a versatile tool that can enhance your daily routines and boost productivity. By leveraging its advanced features, you can streamline tasks, stay organized, and make the most of your time. In this chapter, we'll explore how to use the Echo Spot to set alarms, reminders, and calendar events, utilize Alexa for productivity tasks, integrate with third-party productivity apps, and look at case studies of improved productivity using the Echo Spot.

Using the Echo Spot to Enhance Daily Routines

The Echo Spot 2024 can significantly improve your daily routines by automating tasks and providing timely information. Here are some ways to enhance your day-to-day life with the Echo Spot:

Morning Routine:

Start your day on the right foot with a personalized morning routine. Set up the Echo Spot to wake you up with your favorite music, provide a weather update, and give a summary of your calendar events for the day. You can even have Alexa read the news headlines while you get ready.

Evening Routine:
Wind down in the evening with a relaxing routine. Use the Echo Spot to turn off lights, play calming music or white noise, and adjust the thermostat to your preferred sleeping temperature. Set a reminder to prepare for the next day, such as laying out clothes or packing a lunch.

Meal Prep Routine:
Streamline meal preparation with the Echo Spot. Set timers for cooking, ask Alexa for recipes, and create a shopping list that syncs with your smartphone. You can also use the Echo Spot to control smart kitchen appliances, like turning on a coffee maker or preheating the oven.

Exercise Routine:
Enhance your workout routine with the Echo Spot. Use it to play workout music, set timers for exercise intervals, and track your fitness progress. You can also ask Alexa for workout recommendations or guided meditation sessions.

Setting Alarms, Reminders, and Calendar Events

The Echo Spot 2024 excels at helping you stay on top of your schedule with alarms, reminders, and calendar integration. Here's how to make the most of these features:

Setting Alarms:
To set an alarm, simply say, "Alexa, set an alarm for [time]." You can also customize the alarm sound and set recurring alarms for specific days. For example, "Alexa, set a weekday alarm for 7:00 AM."

Setting Reminders:
Use reminders to keep track of tasks and important events. Say, "Alexa, remind me to [task] at [time]." For example, "Alexa, remind me to call John at 3:00 PM." Alexa will notify you at the specified time with a reminder on the Echo Spot's screen and a voice alert.

Calendar Integration:
Integrate your calendar with Alexa to receive updates on your schedule. You can link Google Calendar, Microsoft Outlook, Apple Calendar, and other popular calendar services. Once linked, you can ask Alexa, "What's on my calendar today?" or "Add an event to my calendar." Alexa can also send notifications for upcoming events, ensuring you never miss an appointment.

To-Do Lists:
Create and manage to-do lists with the Echo Spot. Say, "Alexa, add [task] to my to-do list." You can review your list by asking, "Alexa, what's on my to-do list?" This feature is particularly useful for keeping track of errands, work tasks, and personal goals.

Utilizing Alexa for Productivity Tasks

Alexa is equipped with a range of features that can enhance your productivity. Here are some ways to leverage these capabilities:

Timers and Stopwatch:
Set timers and use the stopwatch function for time management. This is especially useful for tasks that require focus, such as the Pomodoro technique (25 minutes of focused work followed by a 5-minute break).

Notes and Lists:
Create notes and lists to organize your thoughts and tasks. Say, "Alexa, take a note" or "Alexa, create a shopping list." These notes and lists can be accessed through the Alexa app, making it easy to keep track of important information.

Voice Commands for Quick Actions:

Use voice commands to perform quick actions without interrupting your workflow. For example, you can ask Alexa to send an email, make a call, or provide a definition. This hands-free convenience allows you to multitask efficiently.

Daily Briefings:
Set up daily briefings to receive a summary of important information at a specific time each day. This can include news updates, weather forecasts, calendar events, and reminders. Say, "Alexa, what's my flash briefing?" to hear your customized update.

Translation and Language Learning:
Use Alexa for translation and language learning. Ask for translations of words or phrases, or use language learning skills to practice a new language. For example, "Alexa, how do you say 'good morning' in Spanish?"

Integrating with Third-Party Productivity Apps

The Echo Spot 2024 can be integrated with various third-party productivity apps to further enhance its capabilities. Here are some popular apps and how to connect them:

Todoist:

Todoist is a popular task management app that integrates seamlessly with Alexa. To connect Todoist, open the Alexa app, go to "Skills & Games," and search for Todoist. Enable the skill and link your Todoist account. Once connected, you can add tasks, check your to-do list, and manage projects using voice commands.

Trello:
Trello is a project management tool that helps you organize tasks and collaborate with others. To integrate Trello with Alexa, enable the Trello skill in the Alexa app and link your account. You can then use voice commands to add cards, move tasks, and check your boards.

Evernote:
Evernote is a versatile note-taking app that can be linked to Alexa for easy access to your notes and notebooks. Enable the Evernote skill in the Alexa app and connect your account. You can create, search, and update notes using voice commands, making it easy to capture ideas and stay organized.

Google Calendar:
Integrate Google Calendar with Alexa to manage your schedule more efficiently. Link your Google Calendar account through the Alexa app and use voice commands to add events, check your agenda, and receive notifications for upcoming appointments.

Microsoft To Do:
Microsoft To Do is a task management app that syncs with Alexa. Enable the Microsoft To Do skill in the Alexa app and link your account. You can add tasks, view your lists, and mark items as complete using voice commands.

Case Studies of Improved Productivity Using Echo Spot

To illustrate the practical benefits of using the Echo Spot 2024 for productivity, here are some real-world case studies:

Case Study 1: Sarah's Home Office Efficiency

Sarah works from home and uses the Echo Spot 2024 to streamline her workday. She sets up a morning routine that includes a weather update, news briefing, and a summary of her calendar events. Throughout the day, Sarah uses voice commands to set timers for focused work sessions, add tasks to her to-do list, and schedule meetings. By integrating Todoist and Trello, she manages her projects more effectively and collaborates with her team. Sarah's productivity has significantly improved, allowing her to balance work and personal life more efficiently.

Case Study 2: John's Fitness and Wellness Routine

John is committed to maintaining a healthy lifestyle and uses the Echo Spot 2024 to support his fitness and wellness goals. He sets exercise reminders and uses the device to play workout music and guided meditation sessions. John tracks his meals and hydration with the help of third-party apps like MyFitnessPal, which sync with Alexa. By setting up a bedtime routine, John ensures he gets enough rest and wakes up feeling refreshed. The Echo Spot has become an integral part of John's daily wellness routine, helping him stay on track and motivated.

Case Study 3: Lisa's Family Organization

Lisa manages a busy household with two kids and uses the Echo Spot 2024 to keep everyone organized. She sets up routines for morning and bedtime, ensuring the kids wake up on time and go to bed without fuss. Lisa uses Alexa to create shopping lists, set reminders for school events, and schedule family activities. By linking her Google Calendar, Lisa receives notifications for appointments and can easily manage the family's schedule. The Echo Spot has simplified Lisa's life, making it easier to juggle multiple responsibilities and keep the household running smoothly.

Case Study 4: Mark's Project Management

Mark is a project manager who uses the Echo Spot 2024 to stay organized and on top of his tasks. He integrates Trello and Microsoft To Do with Alexa, allowing him to manage projects and tasks using voice commands. Mark sets up routines to receive daily briefings, check project updates, and send reminders to his team. The Echo Spot helps Mark prioritize tasks, meet deadlines, and communicate more effectively with his team. As a result, Mark's productivity has increased, and he feels more in control of his workload.

Case Study 5: Emma's Student Life

Emma is a university student who uses the Echo Spot 2024 to manage her studies and personal life. She sets alarms and reminders for classes, study sessions, and assignments. Emma uses Alexa to create to-do lists and take notes during lectures. By integrating Evernote and Google Calendar, Emma organizes her study materials and schedules exams and deadlines. The Echo Spot helps Emma stay focused and organized, improving her academic performance and reducing stress.

Tips for Maximizing Productivity with the Echo Spot

To make the most of your Echo Spot 2024 for productivity, consider these tips:

1. Personalize Routines: Customize routines to fit your specific needs and preferences. Experiment with different actions and triggers to find what works best for you.

2. Leverage Voice Commands: Use voice commands to perform quick actions and access information without interrupting your workflow. Practice using Alexa to become more comfortable with voice interactions.

3. Integrate Apps: Take advantage of third-party app integrations to extend the functionality of your Echo Spot. Explore the Alexa Skills Store to discover new productivity tools and services.

4. Set Up Daily Briefings: Configure daily briefings to receive a summary of important information at a specific time each day. This can help you start your day informed and prepared.

5. Use Timers and Reminders: Set timers and reminders to stay on track and manage your time effectively. Break tasks into manageable chunks and set reminders for important deadlines.

6. Explore New Skills: Regularly check for new skills and updates in the Alexa app. Enabling new skills can enhance your Echo Spot's capabilities and provide additional productivity features.

7. Stay Organized: Keep your Echo Spot's settings and routines organized. Periodically review and update your routines, alarms, and reminders to ensure they remain relevant and effective.

The Echo Spot 2024 is a powerful tool that can transform the way you manage your daily routines and boost productivity. By leveraging its advanced features and integrating it with third-party productivity apps, you can create a more organized, efficient, and enjoyable lifestyle. Whether you're working from home, managing a household, or pursuing personal goals, the Echo Spot 2024 offers a versatile and user-friendly platform to help you achieve your objectives.

CHAPTER 7

ENTERTAINMENT AND MEDIA

The Echo Spot 2024 is not just a productivity tool; it's also a versatile entertainment hub that can transform the way you enjoy music, audiobooks, and other media. This chapter explores how to use the Echo Spot for entertainment, including its compatibility with streaming services, media control via the touchscreen, unique features like Kindle read-aloud, and tips for personalizing your entertainment settings for an optimal experience.

Using Echo Spot for Music and Audiobooks

The Echo Spot 2024 is designed to deliver high-quality audio for both music and audiobooks. With its front-firing speaker and advanced audio processing, it ensures a rich and immersive listening experience.

Playing Music:
To play music on your Echo Spot, simply use voice commands. You can ask Alexa to play songs, albums,

artists, or genres from your favorite streaming services. For example:
- "Alexa, play some jazz."
- "Alexa, play the latest album by [artist]."
- "Alexa, play my workout playlist."

Music Services:
The Echo Spot supports a variety of music streaming services, including:
- Amazon Music: The default music service for Alexa devices. You can access millions of songs and curated playlists.
- Spotify: Link your Spotify account to enjoy your personalized playlists and discover new music.
- Apple Music: Connect your Apple Music account to play your library and explore Apple's extensive music catalog.
- Pandora: Stream personalized radio stations and discover new artists.
- Tidal: Enjoy high-fidelity sound quality with Tidal's streaming service.

Setting Up Music Services:
To set up a music service, follow these steps:
1. Open the Alexa app on your smartphone or tablet.
2. Tap "More" in the bottom right corner.
3. Select "Settings," then "Music & Podcasts."
4. Tap "Link New Service" and choose the service you want to add.

5. Follow the on-screen instructions to link your account.

Listening to Audiobooks:
The Echo Spot is perfect for audiobook lovers. You can access your Audible library and listen to books with just a voice command. For example:
- "Alexa, play my audiobook."
- "Alexa, read [book title] from Audible."

Controlling Audiobooks:
You can control playback with voice commands or the touchscreen. For example:
- "Alexa, pause the audiobook."
- "Alexa, go back 30 seconds."
- "Alexa, increase the narration speed."

Streaming Services Compatibility and Setup

The Echo Spot 2024 is compatible with a wide range of streaming services, ensuring you have access to a vast library of content for your entertainment needs. Here's how to set up and use some popular streaming services:

Amazon Prime Video:
Although the Echo Spot's screen is small, it's still capable of streaming video content from Amazon Prime Video. To watch videos, simply say:
- "Alexa, play [movie or TV show] on Prime Video."

Netflix:
Stream your favorite shows and movies from Netflix on your Echo Spot. First, link your Netflix account through the Alexa app:
1. Open the Alexa app and go to "Settings."
2. Select "TV & Video," then "Netflix."
3. Tap "Link Your Account" and follow the prompts.

YouTube:
Access YouTube through the Silk or Firefox browser on your Echo Spot. Simply say:
- "Alexa, open YouTube."

Other Streaming Services:
The Echo Spot also supports services like Hulu, Disney+, and HBO Max. To link these services, follow similar steps in the Alexa app under "TV & Video."

Using the Touchscreen for Media Control

The Echo Spot's touchscreen enhances your media control, providing a visual interface that complements voice commands. Here's how to use the touchscreen for various media controls:

Playing and Pausing:

Tap the play/pause button on the screen to control playback without using voice commands.

Skipping Tracks:
Swipe left or right on the screen to skip to the previous or next track when listening to music.

Volume Control:
Adjust the volume by tapping the volume icons or using the on-screen slider.

Browsing Content:
Use the touchscreen to browse through playlists, albums, and genres. Tap on your selection to start playback.

Interacting with Videos:
When watching videos, you can use the touchscreen to pause, play, skip, or rewind. Tap on the screen to bring up playback controls.

Overview of Kindle Read-Aloud and Other Unique Media Features

The Echo Spot 2024 offers unique media features that enhance your entertainment experience, such as the Kindle read-aloud function and more.

Kindle Read-Aloud:

One of the standout features of the Echo Spot is its ability to read Kindle books aloud. This is particularly useful for multitasking or enjoying a book while relaxing. Here's how to use this feature:
- Say, "Alexa, read [book title] from Kindle."
- The Echo Spot will start reading the book aloud, using text-to-speech technology.

Controlling Read-Aloud:
You can control the read-aloud feature with voice commands or the touchscreen. For example:
- "Alexa, pause reading."
- "Alexa, go to the next chapter."
- "Alexa, adjust the reading speed."

Personalizing Your Reading Experience:
Customize your reading experience by adjusting the voice speed and choosing a preferred voice for narration through the Alexa app.

Other Unique Media Features:
- Flash Briefings: Set up flash briefings to get quick updates on news, weather, and other interests. Customize your briefings in the Alexa app by selecting sources and topics.
- Podcasts: Listen to your favorite podcasts by linking podcast services like Apple Podcasts or TuneIn. Say, "Alexa, play [podcast name]."

- Radio Stations: Stream radio stations from around the world using services like iHeartRadio. Say, "Alexa, play [radio station]."

Personalizing Entertainment Settings for an Optimal Experience

To get the most out of your Echo Spot 2024, personalize your entertainment settings to suit your preferences. Here are some tips:

Customizing Music Settings:
- Default Music Service: Set your preferred music service as the default in the Alexa app under "Music & Podcasts."
- Equalizer Settings: Adjust the bass, midrange, and treble settings for optimal sound quality. Say, "Alexa, open equalizer settings" or adjust them in the Alexa app.

Setting Up Playlists:
- Create Playlists: Use your music streaming service to create playlists for different moods and activities. For example, a workout playlist, a relaxing playlist, or a party playlist.
- Voice Commands: Easily access your playlists by saying, "Alexa, play my [playlist name] playlist."

Optimizing Audiobook Settings:
- Narration Speed: Adjust the narration speed to your liking in the Alexa app or by saying, "Alexa, increase/decrease narration speed."
- Bookmarks: Use voice commands to place bookmarks in your audiobooks. Say, "Alexa, add a bookmark here," and "Alexa, go to my last bookmark."

Managing Video Settings:
- Video Quality: Ensure your streaming services are set to the best video quality available, based on your internet connection.
- Subtitles: Enable or disable subtitles through the video service settings or by using voice commands when supported.

Creating Multi-Room Music Groups:
- Setup Multi-Room Music: In the Alexa app, create groups to play music across multiple Echo devices simultaneously. Go to "Devices," select "Multi-Room Music," and follow the prompts to set up your groups.

Exploring New Skills:
- Alexa Skills: Explore and enable new Alexa skills related to entertainment in the Alexa Skills Store. These skills can enhance your Echo Spot's

functionality and provide additional content options.

Using Profiles for Personalized Content:
- Voice Profiles: Set up voice profiles for each household member to get personalized music recommendations and tailored content. Go to "Settings" in the Alexa app, select "Alexa Account," and then "Recognized Voices" to set up voice profiles.

The Echo Spot 2024 offers a rich and diverse entertainment experience, making it a valuable addition to any home. Whether you're streaming music, listening to audiobooks, watching videos, or enjoying unique media features like Kindle read-aloud, the Echo Spot provides a seamless and enjoyable experience. By setting up compatible streaming services, using the touchscreen for intuitive control, and personalizing your settings, you can optimize your entertainment experience to match your preferences. Explore the vast array of entertainment options available and make the Echo Spot 2024 your go-to device for relaxation and enjoyment.

CHAPTER 8

PRIVACY AND SECURITY

As the Echo Spot 2024 becomes an integral part of your smart home ecosystem, it's crucial to understand the privacy and security implications of using such a device. This chapter explores the privacy concerns associated with smart devices, the built-in privacy features of the Echo Spot, and best practices for maintaining security and protecting your data. By being informed and proactive, you can enjoy the benefits of your Echo Spot while ensuring your personal information remains safe and secure.

Understanding Privacy Concerns with Smart Devices

Smart devices like the Echo Spot offer incredible convenience and functionality, but they also raise valid privacy concerns. These concerns often stem from the fact that these devices are always listening for their wake word, storing voice recordings, and potentially capturing video data. Understanding these concerns helps you make informed decisions about how to use your Echo Spot securely.

Key Privacy Concerns:

1. Always Listening: Smart speakers are designed to always listen for their wake word ("Alexa" in the case of Echo devices). This means the device is continuously monitoring ambient sounds, which raises concerns about unintentional recordings.

2. Data Storage: Voice commands and interactions with your Echo Spot are stored on Amazon's servers. While this data is used to improve Alexa's functionality, it also means personal information is being stored remotely.

3. Camera Usage: Devices with built-in cameras, like the Echo Spot, can potentially capture video data. This is particularly concerning if the device is located in private areas like bedrooms.

4. Data Sharing: The integration of third-party skills and services raises concerns about data sharing and how external entities handle your information.

Built-in Privacy Features of the Echo Spot

Amazon has incorporated several privacy features into the Echo Spot to address these concerns and give users control over their data and privacy.

Physical Microphone and Camera Controls:

The Echo Spot includes physical controls to disable the microphone and camera. Pressing the microphone/camera off button on the device instantly disconnects both, ensuring that no audio or video is captured. When the microphone and camera are off, a red indicator light appears on the device, providing a clear visual confirmation.

Voice Command Management:
Users can manage their voice recordings directly through the Alexa app. This includes reviewing, deleting, and setting preferences for how long recordings are stored. For instance, you can configure Alexa to automatically delete recordings after three months or eighteen months.

Privacy Dashboard:
The Alexa app features a Privacy Dashboard where you can manage various privacy settings. This includes controlling how your data is used, managing voice recordings, and adjusting skill permissions. The dashboard provides a centralized location for all privacy-related settings, making it easier to maintain control over your information.

Voice Command for Privacy Settings:
You can use voice commands to manage privacy settings conveniently. For example, saying "Alexa, delete what I just said" or "Alexa, delete everything I said today"

allows you to quickly remove specific recordings without navigating through the app.

Managing Microphone and Camera Settings

Controlling the microphone and camera settings on your Echo Spot is essential for maintaining privacy. Here's how to manage these settings effectively:

Microphone Settings:

1. Disable Microphone: Press the microphone/camera off button on the top of the Echo Spot to disable the microphone. The red light will indicate that the microphone is off.

2. Voice Recording Management: Open the Alexa app, go to Settings > Alexa Privacy > Review Voice History. Here, you can view and delete voice recordings.

3. Mute During Conversations: If you have privacy concerns during sensitive conversations, consider muting the Echo Spot temporarily by pressing the microphone/camera off button.

Camera Settings:

1. Disable Camera: Similar to the microphone, pressing the microphone/camera off button also disables the camera. This is especially useful in private areas like bedrooms.

2. Manage Camera Permissions: In the Alexa app, go to Settings > Device Settings, select your Echo Spot, and adjust camera permissions. You can control which skills and apps have access to the camera.

3. Cover the Camera: For an extra layer of privacy, consider using a physical cover or sticker to block the camera when not in use.

Ensuring Secure Connections and Data Protection

Ensuring that your Echo Spot is connected securely and your data is protected is crucial for maintaining privacy. Here are some steps to enhance the security of your device and data:

Secure Wi-Fi Connection:

1. Use Strong Passwords: Ensure your Wi-Fi network is protected with a strong, unique password to prevent unauthorized access.

2. Enable Encryption: Use WPA3 encryption for your Wi-Fi network if available. WPA2 is also acceptable, but WPA3 offers enhanced security features.

3. Separate Networks: Consider setting up a separate network for your smart devices. This can help isolate them from other devices on your primary network, reducing the risk of cross-device attacks.

Firmware and Software Updates:

1. Keep Firmware Updated: Regularly check for and install firmware updates for your Echo Spot. These updates often include security patches and improvements.

2. Update the Alexa App: Ensure the Alexa app on your smartphone or tablet is always up to date to benefit from the latest security features and enhancements.

Password Management:

1. Strong Account Password: Use a strong, unique password for your Amazon account. Avoid using easily guessable passwords or reusing passwords from other accounts.

2. Two-Factor Authentication: Enable two-factor authentication (2FA) for your Amazon account. This adds an extra layer of security by requiring a second form of verification in addition to your password.

Skill and App Permissions:

1. Review Permissions: Regularly review the permissions granted to Alexa skills and third-party apps. Ensure that only necessary permissions are granted and revoke access for any skills or apps you no longer use.

2. Disable Unused Skills: Disable any skills that you no longer use to reduce potential security risks. In the Alexa app, go to Settings > Skills & Games, and disable or delete unused skills.

Best Practices for Maintaining Privacy and Security

Maintaining privacy and security with your Echo Spot requires ongoing vigilance and proactive management. Here are some best practices to follow:

Be Mindful of Conversations:
1. Sensitive Conversations: Mute the Echo Spot or move it to another room during sensitive conversations to prevent unintended recordings.
2. Awareness of Wake Word: Be aware that the device is always listening for the wake word. Avoid discussing highly sensitive information near the device.

Regular Privacy Audits:
1. Review Voice Recordings: Periodically review and delete voice recordings in the Alexa app. This helps minimize the amount of stored data.
2. Check Privacy Settings: Regularly review and update privacy settings in the Alexa app to ensure they align with your preferences.

Educate Household Members:
1. Inform Family and Guests: Educate family members and guests about the presence of the Echo Spot and its capabilities. Make them aware of how to mute the device if they have privacy concerns.

2. Set Up Household Profiles: Create household profiles for different users to personalize experiences and manage privacy settings effectively.

Use of Smart Home Routines:

1. Automate Privacy Modes: Create routines that automatically mute the Echo Spot during specific times or activities, such as when you're on a work call or during family gatherings.

2. Control Smart Home Devices: Ensure that smart home devices connected to the Echo Spot are configured with appropriate privacy settings and permissions.

Stay Informed About Security Updates:

1. Follow Security News: Stay informed about the latest security news and updates related to smart devices and the Echo Spot. This helps you stay ahead of potential threats and vulnerabilities.

2. Subscribe to Alerts: Consider subscribing to security alerts from Amazon and other trusted sources to receive notifications about critical updates and best practices.

Backup and Recovery:

1. Regular Backups: Regularly back up important data associated with your Amazon account and smart home setup. This ensures that you can quickly recover in case of data loss or device failure.

2. Recovery Options: Familiarize yourself with the account recovery options available through Amazon.

This can help you regain access to your account if it's ever compromised.

The Echo Spot 2024 offers a wealth of features that enhance your smart home experience, but it's essential to remain vigilant about privacy and security. By understanding the potential risks, utilizing built-in privacy features, managing microphone and camera settings, ensuring secure connections, and following best practices, you can enjoy the benefits of your Echo Spot while keeping your personal information safe. Privacy and security are ongoing processes, and staying informed and proactive will help you maintain control over your data and protect your smart home environment.

CHAPTER 9

ADVANCED TIPS AND CUSTOMIZATIONS

The Echo Spot 2024 is a powerful and versatile device that offers a range of advanced settings and customization options for power users. This chapter explores how to personalize the display and touch interface, customize Alexa's responses and interactions, use the Echo Spot in unique and creative ways, and integrate it with other Amazon services and devices. These advanced tips will help you get the most out of your Echo Spot and tailor it to fit your specific needs and preferences.

Advanced Settings and Customizations for Power Users

Power users can take advantage of advanced settings and customizations to enhance their Echo Spot experience. Here are some tips and tricks to unlock the full potential of your device:

Developer Options:

For users with technical expertise, exploring developer options can open up additional customization possibilities. These options allow you to experiment with different settings and features that are not typically accessible through the standard interface.

Custom Skills Development:
If you have programming knowledge, consider creating custom Alexa skills to extend the functionality of your Echo Spot. Amazon provides extensive documentation and tools through the Alexa Skills Kit (ASK), enabling you to develop and deploy your own skills.

Advanced Audio Settings:
Enhance your audio experience by tweaking advanced settings such as equalizer adjustments. Use voice commands or the Alexa app to fine-tune bass, midrange, and treble levels to match your listening preferences.

Voice Profile Customization:
Create and manage multiple voice profiles for different users in your household. This allows Alexa to provide personalized responses and recommendations based on individual preferences and usage patterns.

Personalizing the Display and Touch Interface

The Echo Spot's display and touch interface can be personalized to match your style and preferences. Here's how to customize these elements:

Clock Faces:
Choose from a variety of clock faces to personalize the look of your Echo Spot. Access the settings menu by swiping down from the top of the screen, then navigate to Home & Clock > Clock & Photo Display. Select from analog, digital, and themed clock faces.

Photo Slideshows:
Display personal photos as a slideshow on your Echo Spot. Connect your Amazon Photos account or upload images through the Alexa app. Go to Settings > Home & Clock > Clock & Photo Display > Personal Photos to set up your slideshow.

Brightness and Adaptive Display:
Adjust the brightness of the display to suit your environment. Enable adaptive brightness to automatically adjust the screen based on ambient light. Go to Settings > Display > Brightness to make these adjustments.

Touch Gestures:
Familiarize yourself with touch gestures for easier navigation. Swipe left or right to access different screens,

swipe down to open quick settings, and tap to interact with on-screen elements.

Customizing Alexa Responses and Interactions

Alexa's responses and interactions can be customized to provide a more personalized experience. Here are some ways to tailor Alexa to your liking:

Custom Responses:
Create custom responses for specific commands using Alexa Routines. Open the Alexa app, go to More > Routines, and tap the "+" icon to create a new routine. Set a trigger phrase and add a custom response under the "Add action" section.

Adjust Wake Word:
Change the wake word to something other than "Alexa" if you prefer. Options include "Echo," "Computer," and "Amazon." Go to Settings > Device Settings, select your Echo Spot, and choose Wake Word to make this change.

Voice Training:
Improve Alexa's ability to understand your voice by completing the voice training process. Open the Alexa app, go to Settings > Alexa Account > Recognized Voices, and follow the prompts to train Alexa with your voice.

Preferred News Sources:
Customize your flash briefing to include news sources that interest you. Go to Settings > Flash Briefing in the Alexa app, and add or remove news providers to tailor your daily updates.

Language Preferences:
Change Alexa's language or enable multilingual mode to interact with Alexa in different languages. Go to Settings > Device Settings, select your Echo Spot, and choose Language to select your preferred options.

Using the Echo Spot in Unique and Creative Ways

The Echo Spot's versatility allows for unique and creative uses that go beyond typical smart speaker functions. Here are some innovative ways to use your Echo Spot:

Interactive Storytelling:
Engage children with interactive storytelling skills available through Alexa. These skills allow kids to participate in the story by making choices that influence the plot. Enable skills like "Choose Your Own Adventure" for an immersive experience.

Smart Home Control Center:

Transform your Echo Spot into a central control hub for your smart home. Use the touchscreen to manage smart devices, view live feeds from security cameras, and control lighting, thermostats, and more.

Video Conferencing:
Use the Echo Spot for video conferencing through services like Skype. This is particularly useful for staying connected with family, friends, and colleagues. Enable the Skype skill and link your account to start making video calls.

Digital Photo Frame:
When not in use, set your Echo Spot to display a slideshow of your favorite photos, turning it into a digital photo frame. This adds a personal touch to your device and keeps your cherished memories on display.

Fitness Companion:
Incorporate the Echo Spot into your fitness routine by accessing workout skills, guided meditations, and yoga sessions. Use voice commands to start exercises, track progress, and receive motivation.

Educational Tool:
Leverage educational skills to learn new things. Alexa can help with language learning, trivia games, historical facts, and more. Enable relevant skills in the Alexa app to enhance your knowledge on various topics.

Home Automation with IFTTT:
Integrate your Echo Spot with IFTTT (If This Then That) to create advanced automation routines that involve multiple smart devices and services. Visit the IFTTT website, link your Alexa account, and explore applets to customize your automation setup.

Integration with Other Amazon Services and Devices

The Echo Spot seamlessly integrates with a variety of Amazon services and devices, enhancing its functionality and providing a cohesive user experience. Here's how to make the most of these integrations:

Amazon Prime:
Enjoy the benefits of your Amazon Prime membership through the Echo Spot. Access Prime Video, Prime Music, and Prime Reading. Use voice commands to play movies, stream music, or read Kindle books aloud.

Amazon Fire TV:
Control your Fire TV using voice commands through the Echo Spot. Say, "Alexa, play [movie or TV show] on Fire TV," and the content will start playing on your Fire TV device. You can also use Alexa to search for content, control playback, and manage settings.

Amazon Ring:
Integrate your Ring doorbell and security cameras with the Echo Spot. View live feeds and receive motion alerts directly on the Echo Spot's screen. Say, "Alexa, show the front door," to see who's at your door without leaving your seat.

Amazon Shopping:
Use Alexa to manage your shopping lists and place orders on Amazon. Say, "Alexa, add [item] to my shopping list," or "Alexa, reorder [item] from Amazon." Alexa can also track your orders and provide delivery updates.

Amazon Music and Audible:
Stream music from Amazon Music and listen to audiobooks from Audible. Use voice commands to play specific songs, albums, or genres. For audiobooks, say, "Alexa, read [book title] from Audible," to start listening.

Amazon Echo Family:
Expand your smart home ecosystem by adding other Echo devices. Use multi-room music to synchronize audio playback across multiple devices, or use Drop-In to make intercom-style calls between Echo devices in different rooms.

Amazon Smart Plug:

Control devices connected to an Amazon Smart Plug using the Echo Spot. Say, "Alexa, turn on the coffee maker," to start brewing your morning coffee, or "Alexa, turn off the fan," to save energy.

Amazon Alexa Skills:
Explore and enable new skills to extend the capabilities of your Echo Spot. Visit the Alexa Skills Store to discover skills for productivity, entertainment, smart home control, and more. Regularly check for new skills and updates to keep your Echo Spot fresh and functional.

The Echo Spot 2024 offers a wealth of advanced settings and customization options for power users. By personalizing the display and touch interface, customizing Alexa's responses, and using the device in unique and creative ways, you can unlock its full potential. Integration with other Amazon services and devices further enhances the functionality and provides a seamless user experience. Whether you're a tech enthusiast looking to explore advanced features or a casual user seeking to personalize your device, the Echo Spot offers something for everyone. Embrace these tips and customizations to make the Echo Spot an indispensable part of your smart home ecosystem.

CHAPTER 10

TROUBLESHOOTING AND SUPPORT

Even with the best technology, you may encounter issues with your Echo Spot 2024 from time to time. This chapter provides guidance on resolving common issues, step-by-step troubleshooting guides, and information on when and how to contact Amazon support. Additionally, we'll cover how to access community resources and forums for additional help, as well as tips for maintaining and updating your Echo Spot.

Common Issues and How to Resolve Them

Understanding the common issues that users face with the Echo Spot and knowing how to resolve them can save you time and frustration. Here are some of the most frequent problems and their solutions:

1. **Echo Spot Not Responding to Voice Commands**
   - Solution: Ensure that the device is plugged in and powered on. Check if the microphone is muted (indicated by a red light on the device). If the

device still doesn't respond, try restarting it by unplugging and plugging it back in.

2. Wi-Fi Connectivity Issues
- Solution: Check if your Wi-Fi network is functioning properly and that other devices can connect. Ensure the Echo Spot is within range of the router. Restart both the Echo Spot and the router. If the issue persists, go to the Alexa app and reconnect the Echo Spot to the Wi-Fi network.

3. Poor Sound Quality
- Solution: Adjust the equalizer settings in the Alexa app to improve sound quality. Make sure the Echo Spot is placed on a stable surface away from walls or objects that could cause sound distortion.

4. Problems with Bluetooth Pairing
- Solution: Ensure that the Echo Spot and the Bluetooth device are in pairing mode. In the Alexa app, go to Settings > Device Settings > [Your Echo Spot] > Bluetooth Devices, and select the device you want to pair. If pairing fails, restart both devices and try again.

5. Touchscreen Not Responding

- Solution: Clean the screen with a soft, dry cloth to remove any dirt or fingerprints. Restart the device. If the issue persists, perform a factory reset (noting that this will erase all settings and data).

Step-by-Step Troubleshooting Guides

For more complex issues, follow these detailed troubleshooting guides:

1. Reconnecting to Wi-Fi:
1. Open the Alexa app on your smartphone or tablet.
2. Tap the "Devices" icon at the bottom of the screen.
3. Select "Echo & Alexa," then choose your Echo Spot from the list.
4. Tap "Change" next to the Wi-Fi network.
5. Follow the on-screen instructions to reconnect your Echo Spot to the Wi-Fi network.

2. Factory Resetting Your Echo Spot:
1. Swipe down from the top of the Echo Spot screen and tap the gear icon to open Settings.
2. Scroll down and select "Device Options."
3. Tap "Reset to Factory Defaults."
4. Confirm your selection and wait for the device to reset. This will erase all your settings and data, so use this option as a last resort.

3. Updating the Software:
1. Ensure your Echo Spot is connected to Wi-Fi.
2. Swipe down from the top of the screen and tap the gear icon to open Settings.
3. Scroll down and select "Device Options."
4. Tap "Check for Software Updates." The device will download and install any available updates.

When and How to Contact Amazon Support

If you encounter an issue that you cannot resolve through troubleshooting, it may be time to contact Amazon support. Here's how to do it:

When to Contact Support:
- The device does not power on despite being plugged in.
- You experience persistent connectivity issues that are not resolved by restarting or resetting.
- The touchscreen remains unresponsive even after cleaning and restarting.
- Any other hardware or software issues that you cannot resolve.

How to Contact Amazon Support:
1. Visit the Amazon Customer Service page on the website or open the Amazon app.

2. Navigate to the "Help" section and select "Devices" from the list of categories.
3. Choose "Echo & Alexa," then select your Echo Spot model.
4. Click on "Contact Us" and choose the issue you are experiencing.
5. You can opt to chat with a support representative, request a callback, or browse the FAQs for more information.

Accessing Community Resources and Forums for Additional Help

The Amazon Echo community is vast and active, providing a wealth of knowledge and support. Here's how to access these resources:

Amazon Forums:
Visit the Amazon Devices and Digital Services forum to ask questions, share experiences, and get advice from other Echo Spot users. The forums are moderated by Amazon staff who can also provide official support.

Reddit:
Join the r/amazonecho subreddit to engage with a community of Echo device users. This subreddit is a great place to find tips, solutions to common problems, and discussions about new features and updates.

Facebook Groups:
Search for Facebook groups dedicated to Amazon Echo devices. These groups often have active discussions, troubleshooting tips, and user-generated content that can be helpful.

YouTube:
Many tech enthusiasts and experts create video tutorials and troubleshooting guides for Echo devices. Search YouTube for solutions to specific issues, step-by-step setup guides, and reviews of new features.

Maintaining and Updating Your Echo Spot

Regular maintenance and updates can help ensure your Echo Spot operates smoothly and efficiently. Here are some tips for maintaining your device:

1. Keep Your Echo Spot Clean:
Regularly clean the device with a soft, dry cloth to remove dust and fingerprints. Avoid using harsh chemicals or abrasive materials that could damage the screen or exterior.

2. Regularly Check for Software Updates:
Ensure your Echo Spot is always running the latest software version by regularly checking for updates.

Software updates often include performance improvements, new features, and security patches.

3. Manage Device Storage:
If your Echo Spot supports storing content, periodically check the device storage in the Alexa app and remove unnecessary files or apps to free up space.

4. Optimize Network Connectivity:
To maintain a stable Wi-Fi connection, place your Echo Spot in a location with strong signal strength. Avoid placing the device near potential sources of interference, such as microwaves or cordless phones.

5. Review and Update Skills:
Regularly review the Alexa skills you have enabled and disable any that you no longer use. This helps keep your Echo Spot running efficiently and reduces potential security risks.

6. Monitor and Adjust Settings:
Periodically review and adjust your Echo Spot's settings to ensure they align with your current preferences and usage patterns. This includes privacy settings, notification preferences, and device-specific configurations.

Troubleshooting and support are essential aspects of using any smart device, and the Echo Spot 2024 is no exception. By understanding common issues and knowing how to resolve them, you can keep your device running smoothly. Utilizing step-by-step troubleshooting guides, knowing when and how to contact Amazon support, and accessing community resources can provide additional help when needed. Regular maintenance and updates are crucial for ensuring the longevity and optimal performance of your Echo Spot. With these tips and resources, you can confidently manage any challenges and enjoy the full range of features that your Echo Spot 2024 has to offer.

CHAPTER II

FUTURE OF SMART HOME TECHNOLOGY

The smart home landscape is rapidly evolving, with continuous advancements in technology transforming the way we interact with our living spaces. This chapter explores the emerging trends in smart home technology, future developments in Amazon Echo devices, the role of the Echo Spot in the evolving smart home ecosystem, and predictions for the next generation of smart speakers.

Emerging Trends in Smart Home Technology

The smart home market is witnessing several exciting trends that promise to make our homes more intelligent, efficient, and secure. Here are some of the key trends to watch:

1. Artificial Intelligence and Machine Learning:
AI and machine learning are at the forefront of smart home innovation. These technologies enable devices to learn from user behavior, making predictions and automating tasks with increasing accuracy. For example,

smart thermostats can learn your schedule and adjust temperatures accordingly, while smart speakers like the Echo Spot can offer personalized recommendations based on your preferences.

2. Interoperability and Standardization:
One of the biggest challenges in smart home technology has been the lack of interoperability between devices from different manufacturers. The introduction of standards like Matter (previously known as Project CHIP) aims to address this issue by ensuring seamless communication and compatibility across various smart home devices. This will simplify the setup and integration process for users, making it easier to create cohesive smart home ecosystems.

3. Enhanced Security and Privacy:
As smart home devices become more integrated into our daily lives, security and privacy concerns are becoming increasingly important. Manufacturers are focusing on enhancing security features, such as end-to-end encryption and multi-factor authentication, to protect user data. Additionally, there is a growing emphasis on providing users with greater control over their data and privacy settings.

4. Energy Efficiency and Sustainability:
Smart home technology is playing a crucial role in promoting energy efficiency and sustainability. Devices

like smart thermostats, lighting systems, and energy monitors help users reduce energy consumption and lower their carbon footprint. The integration of renewable energy sources, such as solar panels, with smart home systems is also becoming more common.

5. Voice Control and Natural Language Processing:
Voice control remains a central feature of smart home devices, and advancements in natural language processing are making interactions with voice assistants more intuitive and seamless. Smart speakers are becoming better at understanding context and handling complex commands, providing a more natural and user-friendly experience.

Future Developments in Amazon Echo Devices

Amazon Echo devices have been at the forefront of the smart home revolution, and future developments promise to bring even more functionality and convenience to users. Here are some anticipated advancements:

1. Improved AI Capabilities:
Future Echo devices are expected to feature even more advanced AI capabilities, allowing Alexa to understand

and anticipate user needs with greater accuracy. This includes better context awareness, more natural conversations, and enhanced predictive capabilities.

2. Enhanced Integration with Smart Home Devices:

As the Matter standard gains traction, future Echo devices will offer improved integration with a wider range of smart home products. This will enable users to create more cohesive and interconnected smart home environments.

3. Expanded Health and Wellness Features:

Amazon is likely to continue expanding the health and wellness features of Echo devices. This includes capabilities such as sleep tracking, fitness monitoring, and integration with health-related apps and services. Future Echo devices may also incorporate sensors for monitoring vital signs and providing health insights.

4. More Versatile Form Factors:

While the Echo Spot offers a compact and versatile design, future Echo devices may explore even more diverse form factors. This could include wearables, smart displays with larger screens, and devices specifically designed for different areas of the home, such as the kitchen or bathroom.

5. Advanced Home Automation:

Future Echo devices will likely offer more sophisticated home automation features, enabling users to create complex routines and automations with ease. This includes deeper integration with smart home systems, such as security, lighting, and HVAC, allowing for more comprehensive control.

How the Echo Spot Fits into the Evolving Smart Home Landscape

The Echo Spot 2024 is a pivotal device in the smart home landscape, offering a unique blend of functionality, convenience, and versatility. Here's how it fits into the evolving ecosystem:

1. Central Hub for Smart Home Control:
The Echo Spot serves as a central hub for controlling various smart home devices. Its touchscreen interface and voice control capabilities make it easy to manage lighting, thermostats, security systems, and more from a single device.

2. Personalized User Experience:
With features like voice profiles and customizable settings, the Echo Spot provides a highly personalized user experience. This makes it an ideal choice for households with multiple users, each with their own preferences and routines.

3. Integration with Other Echo Devices:
The Echo Spot seamlessly integrates with other Echo devices, enabling multi-room audio, intercom functionality, and synchronized home automation. This creates a cohesive and interconnected smart home environment.

4. Versatility and Adaptability:
The Echo Spot's compact design and versatile functionality make it suitable for various use cases. Whether it's serving as a bedside alarm clock, a kitchen assistant, or a home security monitor, the Echo Spot adapts to the user's needs.

5. Enhanced Entertainment Capabilities:
The Echo Spot's ability to stream music, audiobooks, and videos, along with its compatibility with various entertainment services, makes it a valuable addition to any home entertainment setup. The device's touchscreen provides an interactive and engaging way to access media content.

Predictions and Expectations for the Next Generation of Smart Speakers

Looking ahead, the next generation of smart speakers is expected to bring several exciting advancements and innovations. Here are some predictions and expectations:

1. Greater Emphasis on AI and Machine Learning:

Future smart speakers will leverage AI and machine learning to provide more intuitive and context-aware interactions. This includes improved voice recognition, natural language processing, and the ability to learn and adapt to user preferences over time.

2. Enhanced Audio and Visual Capabilities:

As technology advances, smart speakers will offer superior audio and visual experiences. This includes higher-quality speakers, advanced sound processing, and higher-resolution displays for devices with screens.

3. Increased Focus on Privacy and Security:

With growing concerns about data privacy and security, future smart speakers will incorporate more robust security features. This includes hardware-level encryption, enhanced user authentication, and greater transparency in data usage and storage.

4. Seamless Integration with Smart Home Ecosystems:

Future smart speakers will offer even more seamless integration with a broader range of smart home devices and systems. This includes compatibility with emerging standards like Matter and enhanced interoperability across different platforms and brands.

5. Expansion of Health and Wellness Features:
As smart home technology continues to intersect with health and wellness, future smart speakers will offer more comprehensive health monitoring and wellness features. This includes integration with health apps, fitness tracking, and real-time health insights.

6. More Diverse Form Factors and Use Cases:
The next generation of smart speakers will explore a wider range of form factors and use cases. This includes wearable devices, smart displays tailored for specific rooms, and devices designed for outdoor use.

7. Advanced Home Automation and Routines:
Future smart speakers will enable users to create more complex and sophisticated home automation routines. This includes greater flexibility in setting triggers and actions, deeper integration with smart home systems, and the ability to automate more aspects of daily life.

The future of smart home technology is filled with exciting possibilities, driven by advancements in AI, interoperability, security, and user experience. The Echo Spot 2024 is well-positioned to play a central role in this evolving landscape, offering a versatile and powerful tool for managing and enhancing smart home environments.

As smart speakers continue to evolve, they will become even more integral to our daily lives, providing seamless interactions, personalized experiences, and advanced automation capabilities. Embracing these innovations will allow users to create smarter, more efficient, and more enjoyable living spaces.

CONCLUSION

As we conclude this comprehensive guide to the Echo Spot 2024, it's essential to recap the key points covered in the book, provide final thoughts on maximizing the benefits of your device, and encourage you to continue exploring the vast world of smart home technology. By implementing what you've learned, you can transform your living space into a more connected, efficient, and enjoyable environment.

Recap of Key Points Covered in the Book

Throughout this book, we have explored various aspects of the Echo Spot 2024, from its initial unboxing to advanced tips and troubleshooting. Here's a summary of the key points discussed:

1. Introduction to the Echo Spot:
We started with an overview of smart home technology and the evolution of smart speakers, highlighting the significance of the Echo Spot 2024 in this landscape. We discussed its relevance and the objectives of this book, setting the stage for an in-depth exploration of the device.

2. The Evolution of Smart Speakers:

We delved into the history of Amazon Echo devices, key milestones, and innovations that have shaped the smart speaker market. We examined the transition from simple speakers to multifunctional smart devices, emphasizing the impact of the Echo Spot's predecessors.

3. Unboxing and First Impressions:

This chapter provided a detailed description of the Echo Spot's packaging, a step-by-step unboxing guide, and initial setup process. We also compared the Echo Spot 2024 with previous models and other Amazon devices, highlighting its unique features.

4. Detailed Features and Specifications:

We explored the comprehensive hardware breakdown of the Echo Spot, including touchscreen functionality, speaker quality, connectivity options, Matter controller capabilities, and new features introduced in the 2024 model.

5. Setting Up Your Echo Spot:

This chapter offered a step-by-step guide to setting up your Echo Spot, connecting it to Wi-Fi and Bluetooth, integrating it with the Alexa app, and customizing settings for optimal use. We also provided tips for a seamless setup experience.

6. Smart Home Integration:
We discussed the smart home ecosystem, how to connect the Echo Spot with various smart home devices, and using Alexa for home automation. We covered creating and managing routines for different scenarios and provided real-world examples of smart home integration.

7. Daily Routines and Productivity:
We highlighted how the Echo Spot can enhance daily routines and boost productivity by setting alarms, reminders, and calendar events, utilizing Alexa for productivity tasks, integrating with third-party apps, and sharing case studies of improved productivity.

8. Entertainment and Media:
This chapter explored using the Echo Spot for music and audiobooks, streaming services compatibility and setup, using the touchscreen for media control, Kindle read-aloud and other unique media features, and personalizing entertainment settings for an optimal experience.

9. Privacy and Security:
We addressed privacy concerns with smart devices, the built-in privacy features of the Echo Spot, managing microphone and camera settings, ensuring secure connections and data protection, and best practices for maintaining privacy and security.

10. Advanced Tips and Customizations:
We offered advanced settings and customizations for power users, personalizing the display and touch interface, customizing Alexa responses and interactions, using the Echo Spot in unique and creative ways, and integrating with other Amazon services and devices.

11. Troubleshooting and Support:
This chapter provided guidance on resolving common issues, step-by-step troubleshooting guides, when and how to contact Amazon support, accessing community resources and forums for additional help, and maintaining and updating your Echo Spot.

12. Future of Smart Home Technology:
We explored emerging trends in smart home technology, future developments in Amazon Echo devices, how the Echo Spot fits into the evolving smart home landscape, and predictions for the next generation of smart speakers.

Final Thoughts on Maximizing the Benefits of the Echo Spot

The Echo Spot 2024 is a powerful and versatile device that can significantly enhance your smart home experience. To maximize its benefits, consider the following tips:

Embrace Customization:
Take advantage of the various customization options available for the Echo Spot. Personalize the display, touch interface, and Alexa responses to suit your preferences and lifestyle. Customizing your device can make interactions more enjoyable and efficient.

Explore Alexa Skills:
The Alexa Skills Store offers a wide range of skills that can extend the functionality of your Echo Spot. Regularly explore new skills and enable those that align with your interests and needs. From productivity tools to entertainment options, there's a skill for almost everything.

Leverage Smart Home Integration:
Integrate your Echo Spot with other smart home devices to create a cohesive and interconnected environment. Use the device as a central hub to control lighting, thermostats, security systems, and more. Automate routines to simplify daily tasks and enhance convenience.

Stay Informed and Updated:
Keep your Echo Spot up to date with the latest firmware and software updates. Regularly check for new features

and improvements that can enhance your user experience. Stay informed about emerging trends and advancements in smart home technology.

Utilize Voice Commands:
Voice commands are a key feature of the Echo Spot. Familiarize yourself with the various commands available and use them to streamline tasks and improve efficiency. Experiment with creating custom commands and routines to automate repetitive actions.

Engage with the Community:
Join forums, social media groups, and online communities dedicated to Amazon Echo devices. Engaging with other users can provide valuable insights, tips, and support. Share your experiences and learn from others to make the most of your Echo Spot.

Encouragement to Continue Exploring Smart Home Technology

The world of smart home technology is vast and continually evolving. As you become more familiar with your Echo Spot and other smart devices, continue to explore new possibilities and innovations. Here are some areas to consider:

Smart Home Ecosystems:

Expand your smart home ecosystem by adding new devices and integrating them with your Echo Spot. Explore products from various manufacturers that are compatible with Alexa and enhance your home's functionality.

Emerging Technologies:
Stay updated on emerging technologies such as AI, machine learning, and IoT (Internet of Things). These advancements are driving the future of smart home technology and can provide new opportunities for automation and personalization.

Energy Efficiency:
Consider incorporating smart devices that promote energy efficiency and sustainability. Smart thermostats, lighting systems, and energy monitors can help reduce your carbon footprint and lower utility bills.

Health and Wellness:
Explore smart home devices and applications focused on health and wellness. From fitness trackers to smart sleep monitors, these technologies can support your health goals and improve your overall well-being.

Security and Privacy:
As you expand your smart home, continue to prioritize security and privacy. Stay informed about best practices for protecting your data and securing your devices.

Regularly review and update privacy settings to maintain control over your information.

Now that you have a comprehensive understanding of the Echo Spot 2024 and its capabilities, it's time to implement what you've learned. Here are some actionable steps to get started:

Set Up Your Echo Spot:
If you haven't already, follow the setup guide to get your Echo Spot up and running. Connect it to your Wi-Fi network, integrate it with the Alexa app, and personalize your settings.

Create Custom Routines:
Experiment with creating custom routines to automate daily tasks. Use the Alexa app to set up routines that streamline your morning, evening, and workday activities.

Integrate Smart Home Devices:
Connect your Echo Spot to other smart home devices and create a centralized control hub. Explore the various smart home products compatible with Alexa and enhance your home's automation.

Explore New Skills:
Visit the Alexa Skills Store and enable skills that interest you. From productivity tools to entertainment options,

there's a wide range of skills to enhance your Echo Spot experience.

Join the Community:
Engage with online communities and forums dedicated to Amazon Echo devices. Share your experiences, ask questions, and learn from other users to maximize the benefits of your Echo Spot.

Stay Informed:
Keep up with the latest updates, features, and advancements in smart home technology. Regularly check for software updates and explore new possibilities to enhance your smart home ecosystem.

By taking these steps, you can fully leverage the capabilities of the Echo Spot 2024 and create a smarter, more connected, and more efficient home. Embrace the future of smart home technology and continue to explore the endless possibilities it offers.

APPENDICES

Glossary of Terms

Alexa: Amazon's cloud-based voice service that powers the Echo and other Amazon devices, enabling voice interaction, music playback, smart home control, and more.

Echo Spot: A compact smart speaker with a touchscreen display, part of the Amazon Echo family, designed to perform a variety of tasks using Alexa.

AI (Artificial Intelligence): The simulation of human intelligence processes by machines, especially computer systems, including learning, reasoning, and self-correction.

Smart Home: A residence equipped with devices that automate tasks normally handled by humans, using Internet-connected devices and voice assistants like Alexa.

Wi-Fi: A technology that allows devices to connect to the Internet wirelessly.

Bluetooth: A wireless technology standard for exchanging data over short distances.

Matter: An industry-unifying standard for smart home devices that aims to ensure seamless interoperability between different brands and ecosystems.

Voice Command: A spoken instruction that a smart device, like the Echo Spot, can understand and respond to.

Routines: Predefined sets of actions that Alexa can perform automatically based on triggers like voice commands, schedules, or device interactions.

Firmware: The permanent software programmed into a device, providing the necessary instructions for how the device communicates with other hardware.

IFTTT (If This Then That): A free web-based service that creates chains of simple conditional statements, called applets, to automate interactions between various devices and services.

Frequently Asked Questions (FAQ)

Q: How do I set up my Echo Spot for the first time?

A: Plug in your Echo Spot, follow the on-screen instructions to connect to Wi-Fi, sign in to your Amazon account, and set up your preferences. Refer to Chapter 4 for a detailed setup guide.

Q: Can I change the wake word for my Echo Spot?
A: Yes, you can change the wake word to "Echo," "Computer," or "Amazon." Go to Settings > Device Settings > [Your Echo Spot] > Wake Word in the Alexa app.

Q: How do I connect my Echo Spot to a Bluetooth speaker?
A: Enable Bluetooth on your speaker, then say, "Alexa, pair Bluetooth." Select your speaker from the list of available devices in the Alexa app.

Q: Is it possible to use the Echo Spot as a baby monitor?
A: Yes, you can use the Echo Spot as a baby monitor by setting up a Drop-In feature between two Echo devices.

Q: How do I control smart home devices with my Echo Spot?
A: Ensure your smart home devices are compatible with Alexa and connected to the same Wi-Fi network. Use the Alexa app to discover and add new devices, then control them with voice commands or the touchscreen.

Q: What should I do if my Echo Spot is not responding?
A: Check if the device is plugged in and powered on, make sure the microphone is not muted, and try restarting the device. Refer to Chapter 10 for more troubleshooting steps.

Q: How can I delete my voice recordings?
A: You can delete voice recordings via the Alexa app by going to Settings > Alexa Privacy > Review Voice History. You can also use voice commands like "Alexa, delete what I just said."

Additional Resources for Further Learning

Amazon Help and Support:
- Amazon Echo Help: [Amazon Echo Help](https://www.amazon.com/gp/help/customer/display.html?nodeId=201602060)
- Alexa Help: [Alexa Help](https://www.amazon.com/gp/help/customer/display.html?nodeId=202011830)

Online Communities:
- Amazon Devices and Digital Services Forum: [Amazon Forum](https://www.amazonforum.com/)

- Reddit: [r/amazonecho](https://www.reddit.com/r/amazonecho/)
- Facebook Groups: Search for "Amazon Echo Users" or similar groups on Facebook.

Learning Platforms:
- Udemy: Courses on smart home technology and Alexa skills development.
- Coursera: Technology and programming courses that include IoT and smart home device integration.

Books:
- "The Smart Home Manual: How to Automate Your Home to Keep Your Family Entertained, Comfortable, and Safe" by Marlon Buchanan.
- "Smart Homes for Dummies" by Danny Briere and Pat Hurley.

List of Compatible Devices and Services

Smart Home Devices:
- Lighting: Philips Hue, LIFX, TP-Link Kasa, Sengled
- Thermostats: Ecobee, Nest, Honeywell
- Security Cameras: Ring, Arlo, Blink, Nest
- Door Locks: August, Schlage, Yale

- Plugs: Amazon Smart Plug, TP-Link Kasa Smart Plug, Wemo Mini

Music and Media Services:
- Music: Amazon Music, Spotify, Apple Music, Pandora, Tidal
- Video: Prime Video, Netflix, YouTube, Hulu, Disney+
- Audiobooks: Audible
- Podcasts: Apple Podcasts, TuneIn

Productivity and Utility:
- Task Management: Todoist, Any.do, Microsoft To Do
- Calendars: Google Calendar, Microsoft Outlook, Apple Calendar
- Fitness and Health: Fitbit, MyFitnessPal

Voice Assistants and Ecosystems:
- Voice Assistants: Alexa, Google Assistant, Apple Siri
- Ecosystems: Matter, Zigbee, Z-Wave

Detailed Specifications and Technical Information

General Specifications:
- Device Name: Echo Spot 2024

- Dimensions: 4.1" x 3.8" x 3.6" (104 mm x 97 mm x 91 mm)
- Weight: 405 grams
- Display: 2.83-inch touchscreen, 240x320 pixels
- Audio: 1.73-inch front-firing speaker
- Microphones: Far-field microphone array
- Processor: MediaTek MT8163

Connectivity:
- Wi-Fi: Dual-band Wi-Fi supports 802.11 a/b/g/n (2.4 and 5 GHz) networks
- Bluetooth: Advanced Audio Distribution Profile (A2DP) support for audio streaming
- Matter: Integrated Matter controller for seamless smart home device connectivity

Power:
- Power Adapter: 15W (5V/3A) power adapter included
- Power Consumption: Less than 1W in standby mode

Software and Features:
- Operating System: Alexa Voice Service (AVS)
- Voice Assistant: Alexa with support for thousands of skills
- Security: Multi-factor authentication, end-to-end encryption for voice recordings

- Privacy Features: Microphone/camera off button, physical camera shutter

Supported Languages:
- English, Spanish, French, German, Italian, Japanese, and more

Audio Formats Supported:
- Music: MP3, FLAC, AAC, HE-AAC, HE-AAC v2, Vorbis
- Audiobooks: Audible, AAX

Warranty and Support:
- Warranty: 1-year limited warranty and service included
- Customer Support: 24/7 customer service via phone, chat, and email

By understanding these specifications and the detailed technical information provided, you can make the most informed decisions about integrating the Echo Spot 2024 into your smart home ecosystem.

The Echo Spot 2024 is a versatile and powerful device that can significantly enhance your smart home experience. With its advanced features, customizable settings, and extensive compatibility, it offers endless possibilities for automation, entertainment, productivity, and more. This appendices section provides the necessary resources, FAQs, and detailed information to help you fully utilize your Echo Spot and troubleshoot any issues that may arise.

As you continue to explore and implement the knowledge gained from this book, remember that the smart home landscape is ever-evolving. Stay informed about new developments, engage with the community, and keep experimenting with new setups and integrations to create a smart home environment that best suits your needs and lifestyle.

ABOUT THE AUTHOR

Michael S. Nilsson is a leading expert in smart home technology and an advocate for integrating cutting-edge innovations into everyday living. With a background in electrical engineering and a passion for exploring the latest technological advancements, Michael has dedicated his career to simplifying and enhancing the modern home through intelligent automation.

Michael earned his degree in Electrical Engineering from a prestigious university, where he developed a deep understanding of the principles that underpin modern smart home systems. His early career saw him working with several top technology firms, where he played a crucial role in developing and deploying smart home solutions. Michael's hands-on experience with a wide range of smart devices and platforms has given him a unique perspective on the practical applications and benefits of these technologies.

Michael's interest in smart homes began as a personal quest to improve the efficiency and convenience of his own living space. This personal interest quickly blossomed into a professional passion, leading him to write extensively on the subject. His work focuses on demystifying smart home technology for the average consumer, providing clear, actionable advice that empowers people to make the most of their smart home investments.

As an author, Michael S. Nilsson has penned numerous articles, guides, and books on smart home technology. His writing is characterized by a clear, accessible style that breaks down complex topics into easy-to-understand concepts. Michael's goal is to make smart home technology approachable for everyone, from tech enthusiasts to those who are just beginning their smart home journey.

Michael has contributed to several leading technology publications and websites, where his insights on smart home trends, device reviews, and integration tips are highly valued. His practical advice has helped countless readers enhance their homes with the latest smart devices, creating environments that are not only more efficient but also more enjoyable to live in.

When he's not writing or tinkering with the latest gadgets, Michael enjoys spending time with his family in their smart home, which serves as both a personal sanctuary and a testing ground for new technologies. His home is a living example of the potential of smart home technology, seamlessly blending comfort, convenience, and security.

Michael is also an avid traveler and enjoys exploring new places, always on the lookout for innovative uses of technology in different cultures. His travels often inspire new ideas and perspectives, which he incorporates into his writing.

Looking to the future, Michael plans to continue his mission of educating and inspiring others about the benefits of smart home technology. He is currently working on his next book, which will delve even deeper into advanced smart home integrations and the future of connected living. Michael is also exploring opportunities to collaborate with tech companies and startups to bring even more innovative solutions to the market.

Through his work, Michael S. Nilsson aims to make smart home technology not just a luxury for the tech-savvy but a standard that enhances the quality of life for everyone. His dedication to this vision continues to drive his writing, research, and advocacy in the ever-evolving world of smart home technology.

Made in the USA
Las Vegas, NV
12 October 2024

96731209R00079